SCIENCE

INSTANT REFERENCE

INDIAN PRAIRIE PUBLIC LIBRARY
401 Plainfield Road
Darien, IL 60561

TEACH YOURSELF®

For UK orders: please contact Bookpoint Ltd, 78 Milton Park, Abingdon, Oxon OX14 4TD. Telephone: (44) 01235 400414, Fax: (44) 01235 400454. Lines are open 9.00–6.00, Monday to Saturday, with a 24-hour message answering service.
E-mail address: orders@bookpoint.co.uk

For USA and Canada orders: please contact NTC/Contemporary Publishing, 4255 West Touhy Avenue, Lincolnwood, Illinois 60646–1975, USA. Telephone: (847) 679 5500, Fax: (847) 679 2494.

Long renowned as the authoritative source for self-guided learning — with more than 40 million copies sold worldwide — the *Teach Yourself* series includes over 200 titles in the fields of languages, crafts, hobbies, business, computing and education.

British Library Cataloguing in Publication Data
A catalogue record for this title is available from the British Library.

Library of Congress Catalog Card Number: On file

First published in UK 2000 by Hodder Headline Plc, 338 Euston Road, London NW1 3BH.

First published in US by NTC/Contemporary Publishing, 4255 West Touhy Avenue, Lincolnwood (Chicago), Illinois 60646-1975, USA.

The 'Teach Yourself' name and logo are registered trademarks of Hodder & Stoughton.

Copyright © 2000 Helicon Publishing Ltd

In UK: All rights reserved. No part of this publication may be reproduced or transmitted in any form or by any means, electronic or mechanical, including photocopying, recording, or any information storage or retrieval system, without permission in writing from the publisher or under licence from the Copyright Licensing Agency Limited. Further details of such licences (for reprographic reproduction) may be obtained from the Copyright Licensing Agency Ltd, 90 Tottenham Court Road, London W1P 9HE.

In US: All rights reserved. No part of this publication may be reproduced, stored in a retrieval system, or transmitted in any form or by any means, electronic, mechanical, photocopying, or otherwise, without prior permission of NTC/Contemporary publishing.

Picture credits: Ann Ronan Picture Library; 10, 56, 83, 94, 120, Image Select; 149.

Text editor: Robert Snedden
Typeset by TechType, Abingdon, Oxon
Printed in Great Britain for Hodder & Stoughton Educational, a division of Hodder Headline Plc, 338 Euston Road, London NW1 3BH, by Cox & Wyman Ltd, Reading, Berkshire.

Impression number	10 9 8 7 6 5 4 3 2 1
Year	2006 2005 2004 2003 2002 2001 2000

Contents

A–Z entries	1
Appendix	197
Amino acids	199
Biochemistry: chronology	199
Biology: chronology	201
Chemistry: chronology	203
Cloning: chronology	206
Genetics: chronology	206
Nuclear energy: chronology	212
Physics: chronology	213
Physical constants	216

Bold type in the text indicates a cross reference. A plural, or possessive, is given as the cross reference, i.e. is in bold type, even if the entry to which it refers is singular.

A

absolute zero

The lowest **temperature** possible, according to **kinetic theory**, at which **molecules** are in their lowest **energy** state and stop moving. Absolute zero marks the beginning of the Kelvin temperature scale (named after the Irish physicist William Thomson Kelvin (1824–1907)), zero kelvin (0 K), equivalent to −273.15°C/−459.67°F.

Although the third law of **thermodynamics** indicates the impossibility of reaching absolute zero in practice, physicists at the University of Sussex, UK, currently hold the record for the closest approach, having cooled a **gas** to a few hundred billionths of a degree above absolute zero in 1999. They were investigating a theory of Albert **Einstein's**, developed with Indian physicist Satyendra Bose (1894–1974), that **atoms** cooled close to absolute zero would form a 'superatom'. This peculiar state of matter is known as a Bose–Einstein Condensate. The researchers first pre-cooled a cloud of rubidium atoms by bouncing **lasers** off them. The atoms were then trapped in a strong magnetic field that allowed only the hottest atoms to escape. Those that remained were cold enough to form the condensate.

SUPERFLUID

Near absolute zero, the physical properties of some materials change dramatically; for example, helium becomes a 'superfluid', flowing without **friction**. If you spin a bowl of superfluid helium, the helium stays still while the bowl spins around it!

acid

A compound that is typically a corrosive or sour-tasting liquid. All acids produce hydrogen **ions** (H+) when dissolved in water. For example hydrochloric acid is produced when hydrogen chloride gas (HCl(g)) reacts with water (aq):

$$HCl(g) + aq = H^+(aq) + Cl^-(aq)$$

The strength of an acid is measured by its hydrogen-ion concentration, which is indicated by a **pH** value. All acids have a pH below 7.0. Acids react with **alkalis** to form a **salt** and water; this is called neutralization. For example, hydrochloric acid added to sodium hydroxide gives the salt sodium chloride (common salt) plus water.

$$HCl(aq) + NaOH(aq) = NaCl(aq) + H_2O$$

Strong acids are corrosive; dilute acids have a sour or sharp taste, although in some organic acids this may be partially masked by other flavour characteristics. Acids give a specific colour reaction with indicators; for example, litmus turns red.

Acids can be classified according to their basicity (the number of hydrogen atoms available to react with an alkali, or base) and their degree of ionization (how many of the available hydrogen atoms are ionized in water). Dilute sulphuric acid, for example, is classified as a strong (highly ionized), dibasic (containing two replaceable hydrogen atoms) acid.

acid rain

Rainfall that is acidic, thought to be caused by pollutants such as sulphur dioxide and oxides of nitrogen dissolving in rainwater to form acids. Sulphur dioxide is released into the atmosphere as a result of burning fossil fuels and nitrogen oxides are produced by car exhausts and a variety of industrial processes. Rain can be naturally slightly acidic as carbon dioxide in the atmosphere dissolves to give weak carbonic acid.

Acid rain has been linked with damage to forests and the death of organisms in lakes in places such as Scandinavia and North America. It can also cause damage to buildings by corroding stonework.

acoustics

The science of **sound** and its transmission; specifically the behaviour and characteristics of sound in a particular space, such as a room or theatre. Sound **energy** spreads as vibrations in the form of pressure **waves** that are absorbed by soft objects, such as drapery and people, and reflected as **echoes** by hard surfaces, such as walls and ceilings.

- Echoes that are too distinct make the original sounds difficult to hear.
- Too much sound absorption by drapery and other soft materials deadens the original sounds.

> **PERFORMANCE AND SOUND**
>
> In a well-designed auditorium the echoes bouncing around arrive so frequently at the ear, finally dying down, that the listener registers them merely as a slight extension of the original sound. Auditorium design must also allow the performers to hear each other and special sound reflectors may be erected to achieve this end. Unwanted noise can be avoided by, for example, using carpeting to prevent floor vibration, by soundproofing the building with the aid of double glazing, or masking the outside of the building with planted trees.

adaptation

An adapatation is any change in the structure or function of an organism that allows it to survive and reproduce more effectively in its **environment**. For example, the webbed feet of ducks or otters are adaptations to living in **water**, enabling them to swim more efficiently. In **evolution**, adaptation is thought to occur as a result of random **variation** in the genetic make-up of organisms coupled with **natural selection**. If a **species** cannot adapt to changes in its environment, for example if it cannot cope with sudden temperature changes or the arrival of a new predator, it may become **extinct**.

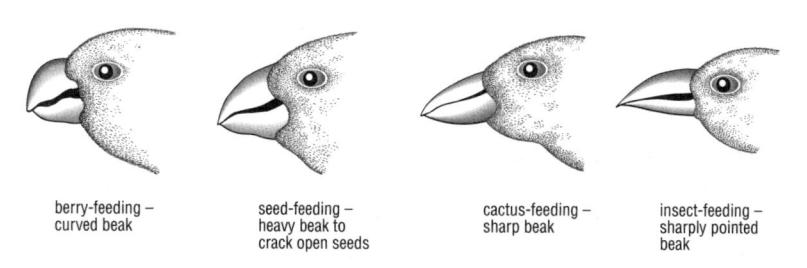

berry-feeding – curved beak

seed-feeding – heavy beak to crack open seeds

cactus-feeding – sharp beak

insect-feeding – sharply pointed beak

adaptation *The different types of beak found in birds are examples of adaptation.*

Physiological adaptation
Adaptation is said to occur in sense organs when the sensitivity of the organ alters in response to changes in environmental conditions, for instance an increase in the size of the eye's pupil to admit more light as night falls.

additive
Any natural or artificial chemical added to alter the colour, texture, or flavour of food, improve its nutritional value, or lengthen its shelf-life. The many chemical additives used are subject to regulation, and food companies in many countries are now required by law to list the additives used in their products. Within the European Union, approved additives are given an official E number. It can be difficult to know how to test the safety of such substances; many natural foods contain toxic substances that could not pass the tests applied today to new products. Individuals may be affected by constant exposure even to traces of certain additives and may suffer side effects ranging from headaches and hyperactivity to **cancer**.

Uses of additives
- *Flavours* increase the appeal of the food by altering or intensifying its taste.
- *Colours* are used to enhance the visual appeal of certain foods. These may be used to restore colour lost in processing and to give a uniform colour to products.
- *Enhancers* are used to increase or reduce the taste and smell of a food without imparting a flavour of their own.
- *Nutrients* replace or enhance food value. Minerals and vitamins are added if the diet might otherwise be deficient or to replace those lost during processing, for example, B vitamins lost in the milling process are restored to cereal products.
- *Preservatives* control natural oxidation and the action of micro-organisms, slowing down the rate of spoilage.
- *Emulsifiers and surfactants* regulate the consistency of fats in the food and on the surface of the food in contact with the air. They modify the texture of food and prevent the ingredients of a mixture from separating out.
- *Thickeners* regulate the consistency of food.
- *Leavening agents* lighten the texture of baked goods without the use of yeasts.

- *Acidulants* sharpen the taste of foods.
- *Bleaching agents* assist in the ageing and whitening of flours.
- *Anticaking agents* prevent powdered products from forming solid lumps.
- *Antioxidants* prevent fatty foods from going rancid by inhibiting their natural oxidation.
- *Humectants* control the humidity of the product by absorbing and retaining moisture.
- *Clarifying agents* are used in fruit juices, vinegars, and other fermented liquids to improve appearance.
- *Firming agents* restore the texture of vegetables that may be damaged during processing.

aerobic

Organisms that require oxygen for the efficient release of **energy** contained in **food molecules** are said to be aerobic. Oxygen is used to convert glucose to carbon dioxide and **water**, thereby releasing energy. Almost all organisms, with the exception of certain **bacteria**, yeasts and internal **parasites**, are aerobic. Aerobic reactions occur inside every **cell** and lead to the formation of energy-rich compounds that are used by the cell to drive its metabolic processes. Aerobic organisms die in the absence of oxygen, but certain organisms and cells, such as those found in muscle tissue, can function for short periods without oxygen. **Anaerobic** organisms can survive without oxygen and in fact are poisoned by it.

aerodynamics

The study of the forces exerted by air or other gases in motion, for example the airflow around aircraft moving at speed through the atmosphere. The acceleration of the air over the rounded leading edge of an aeroplane wing and across the curved upper surface results in a substantial reduction in **pressure** which, combined with a slight increase in pressure on the

SUPERSONIC vs HYPERSONIC

- The most efficient of supersonic (faster than sound) aircraft shapes are slender and thin-winged, with sharp edges and pointed noses.
- The most efficient hypersonic (speeds greater than five times that of sound) shapes are blunt wedges with rounded noses and in some cases no wings at all.

undersurface, lifts the aircraft. The aim is usually to design the aircraft to produce a streamlined flow, with a minimum of turbulence in the moving air. Excessive heating caused by **friction** means that hypersonic vehicles have to fly relatively high; in fact, a safe 'corridor' exists for such vehicles, with a lower boundary determined by structural temperature and an upper boundary set by the available lift.

aerosol
Particles of **liquid** or **solid** suspended in a **gas**. Fog is a common natural example. Aerosol cans contain a substance such as scent or cleaner packed under **pressure** with a device for releasing it as a fine spray.

alcohol
Any member of a group of organic chemical **compounds** with characteristic properties, not least of which is the effect of one group member on the human brain. In an alcohol one of the hydrogen **atoms** in a **hydrocarbon** molecule has been replaced by an oxygen-hydrogen pair (OH). Alcohols may be **liquids** or **solids**, according to the size and complexity of the **molecule**. The main uses of alcohols are as **solvent**s for gums, resins, lacquers, and varnishes; in the making of dyes; for essential oils in perfumery; and for pharmaceuticals. The alcohol produced naturally in the **fermentation** process and consumed as part of alcoholic beverages is called ethanol.

THE FIVE SIMPLEST ALCOHOLS

- methanol or wood spirit (methyl alcohol, CH_3OH)
- ethanol (ethyl alcohol, C_2H_5OH)
- propanol (propyl alcohol, C_3H_7OH)
- butanol (butyl alcohol, C_4H_9OH)
- pentanol (amyl alcohol, $C_5H_{11}OH$).

The lower alcohols are liquids that mix with water; the higher alcohols, such as pentanol, are oily liquids immiscible with water; and the highest are waxy solids – for example, hexadecanol (cetyl alcohol, $C_{16}H_{33}OH$) and melissyl alcohol ($C_{30}H_{61}OH$), which occur in sperm-whale oil and beeswax respectively.

algae
A diverse groups of plants that range in size from single-celled **species** to the seaweeds. Algae were formerly included in the same division of the

plant kingdom as fungi (see: **fungus**) and **bacteria**, neither of which are now considered to be plants at all. They are now classified into a number of divisions including green algae, found in freshwater and on land; stoneworts, found on land and resembling higher plants; golden brown algae; brown algae, which are mainly found in the sea and include kelps, the largest of the algae; the microscopic diatoms and the blue green algae or cyanobacteria. There is debate as to the true **classification** of the blue-green algae. Their **cell** structure is similar to that of bacteria, although they carry out **photosynthesis** just like green plants.

alkali

Alkalis neutralize **acids** and are soapy to the touch. The strength of an alkali is measured by its hydrogen-ion concentration, given by its **pH** value. All alkalis have a pH above 7.0. Their solutions all contain the hydroxide **ion** OH-, which gives them a characteristic set of properties. Alkalis react with acids to form a **salt** and **water** (neutralization). For example, potassium hydroxide and nitric acid gives potassium nitrate and water. They give a specific colour reaction with indicators; for example, litmus turns blue. The hydroxides of **metals** are alkalis. They may be divided into strong and weak alkalis: a strong alkali (for example, potassium hydroxide, KOH) ionizes completely when dissolved in water, whereas a weak alkali (for example, ammonium hydroxide, NH_4OH) exists in a partially ionized state in solution.

THE FOUR MAIN ALKALIS

- sodium hydroxide (caustic soda, NaOH)
- potassium hydroxide (caustic potash, KOH)
- calcium hydroxide (slaked lime or limewater, $Ca(OH)_2$)
- aqueous ammonia ($NH_{3\ (aq)}$).

---

allele

Organisms, such as humans, have two sets of **chromosomes**. Therefore each organism will have two copies of each **gene**, one on each of a pair of chromosomes. Each of the alternative copies of the gene is called an allele. Blue and brown eyes in humans are determined by different alleles of the gene for eye colour, for example.

- *homozygous gene*: one in which both alleles are identical
- *heterozygous gene*: one in which the alleles are different

- *dominant allele*: one that masks the effect of another allele
- *recessive allele*: one that has its effects masked by a dominant allele.

The influence of the dominant allele is the one that is shown in the organism's development. It is said to be expressed. The recessive allele is not expressed. A trait that results from a recessive allele is evident only in an individual that is homozygous for the recessive trait. The blue-eye allele is recessive, for example, therefore blue-eyed individuals must have two blue-eyed alleles.

alloy

An alloy is a **metal** blended with some other metallic or nonmetallic substance to give it special qualities, such as resistance to corrosion, greater hardness, or tensile strength. Alloys are usually made by melting the metals together.

Among the oldest alloys is bronze (mainly an alloy of copper and tin), the widespread use of which ushered in the Bronze Age. Complex alloys are now common; for example, in dentistry, where a cheaper alternative to gold is made of chromium, cobalt, molybdenum, and titanium. Among the most recent alloys are superplastics: alloys that can stretch to double their length at specific **temperatures**, permitting, for example, their injection into moulds as easily as plastic.

- *Master alloys* or foundry alloys are compositions made only for the purpose of melting with other metals to form alloys. They are used to overcome the problems of alloying metals of widely differing melting points.
- *Shape memory alloys* are imprinted with a shape so that even after distortion, a threshold temperature will bring about a return to the original shape. Nitinol, an alloy of titanium and nickel, is an example.
- *Brass* (35–10% zinc, 65–90% copper) is used in decorative metal work, plumbing fittings, industrial tubing.
- *Bronze – common* (2% zinc, 6% tin, 92% copper) is used in machinery and decorative work.
- *Bronze – aluminium* (10% aluminium, 90% copper) is used for machinery castings.
- *Bronze – coinage* (1% zinc, 4% tin, 95% copper) used in coins.

- *Cast iron* (2–4% carbon, 96–98% iron) is used in decorative metalwork, engine blocks, industrial machinery.
- *Dentist's amalgam* (30% copper, 70% mercury) is used for dental fillings.
- *Duralumin* (0.5% magnesium, 0.5% manganese, 5% copper, 95% aluminium) used in framework of aircraft.
- *Gold – coinage* (10% copper, 90% gold) coins.
- *Lead battery plate* (6% antimony, 94% lead) used in car batteries.
- *Manganin* (1.5% nickel, 16% manganese, 82.5% copper) used in resistance wire.
- *Nichrome* (20% chromium, 80% nickel) used for heating elements.
- *Pewter* (20% lead, 80% tin) used for utensils.
- *Silver – coinage* (10% copper, 90% silver) used for coins.
- *Solder* (50% tin, 50% lead) used for joining iron surfaces.
- *Steel – stainless* (8–20% nickel, 10–20% chromium, 60–80% iron) used for kitchen utensils.
- *Steel – armour* (1–4% nickel, 0.5–2% chromium, 95–98% iron) used for armour plating.
- *Steel – tool* (2–4% chromium, 6–7% molybdenum, 90–95% iron) used for tools.

amino acid

These small **molecules** are the building blocks of **proteins**. When two or more amino acids are joined together, they form a peptide; proteins are made up of peptide chains folded or twisted in characteristic shapes. Many different proteins are found in the **cells** of living organisms, but they are all made up of the same 20 amino acids, joined together in varying combinations (although other types of amino acid do occur infrequently in nature). Eight of these, the essential amino acids, cannot be synthesized by humans and must be obtained from the diet. Children need a further two amino acids that are not essential for adults. Other animals also need some amino acids in their diet.

Amino acids are water-soluble organic molecules, mainly composed of **carbon, oxygen**, hydrogen and nitrogen, and containing both a basic amino group (NH_2) and an acidic carboxyl (COOH) group.

> **PLANT AMINO ACIDS**
>
> Green plants can manufacture all the amino acids they need from simpler molecules, using energy from the Sun and minerals (including nitrates) from the soil.

Ampère, Andre Marie (1775–1836)

French physicist and mathematician who made many discoveries in **electromagnetism** and electrodynamics. The unit of **electric current**, the ampere, is named after him.

In a series of papers beginning in 1820 Ampère set out the basic laws of electromagnetism (which he called electrodynamics to differentiate it from the study of stationary electric forces, which he called electrostatics). He showed that two parallel wires carrying current in the same direction attract each other, whereas when the currents are in opposite directions, mutual repulsion results. He also predicted and demonstrated that a helical coil of wire (which he called a solenoid) behaves like a bar magnet while it is carrying an electric current. Trying to explain electromagnetism, Ampère proposed that **magnetism** is merely **electricity** in motion.

Ampère *French physicist and mathematician Andre Ampère.*

amphibian

The name amphibian comes from the Greek meaning 'both lives', referring to a **life cycle** that takes place both in **water** and on land. Amphibians are members of the class Amphibia of **vertebrates** that contains the frogs, toads, newts, salamanders, and the worm-like caecilians. Amphibians generally spend their **larval** (tadpole) stage in fresh water, transferring to land at maturity (after **metamorphosis**) and generally returning to water to breed. Millions of years ago amphibians were the first animals to leave the oceans and move on to the land. Like **fish** and **reptiles**, they continue to grow throughout life, and cannot maintain a temperature greatly differing from that of their **environment**.

There are 4,553 known species of amphibian:

- *frogs and toads*: 4,000 species
- *newts and salamanders*: 390 species
- *caecilians*: 163 species.

> **ENDANGERED AMPHIBIANS**
>
> The 1990s has seen a marked decline in amphibian **populations** worldwide from causes other than loss of **habitat**, notably infectious **diseases**. One of the causes of death is a previously unknown parasitic fungus. The **fungus** was discovered in 1998 independently by US and Australian researchers.

anaerobic

Organisms that do not require oxygen for the release of **energy** from food molecules such as glucose are said to be anaerobic. Anaerobic organisms include many **bacteria**, yeasts, and internal **parasites**. Anaerobic respiration is a primitive and inefficient form of energy release, deriving from the early period of life on Earth when oxygen was missing from the atmosphere. It may also be seen as an **adaptation** to survival in **habitats** such as the muddy bottom of a polluted river where oxygen is in scarce supply.

- *Obligate anaerobes* cannot function in the presence of oxygen and are poisoned by it.
- *Facultative anaerobes* can function with or without oxygen.

Anaerobic organisms release much less of the available energy from their **food** than do **aerobic** organisms. In plants, yeasts, and bacteria, anaerobic **respiration** results in the production of **alcohol** and carbon dioxide, a process that is exploited by both the brewing and the baking industries (see **fermentation**).

Normally aerobic animal **cells** can respire anaerobically for short periods of time when oxygen levels are low, but are ultimately fatigued by the build-up of the lactic **acid** produced in the process. This is seen particularly in muscle cells during intense activity, when the demand for oxygen can outstrip supply.

anaesthetic
A **drug** that produces loss of sensation or consciousness; producing a state of anaesthesia, in which the patient is insensitive to stimuli. An anaesthetic agent acts either by preventing stimuli from being sent (local), or by removing awareness of them (general). Anaesthesia may also happen as a result of nerve disorder.

analgesic
An agent for relieving pain. Pain is felt when electrical stimuli travel along a nerve pathway, from peripheral nerve fibres to the brain via the spinal cord. Analgesic **drugs** act on both. Opiates, such as morphine, alter the perception or appreciation of pain and are effective in controlling 'deep' visceral (internal) pain. Non-opiates, such as aspirin, paracetamol, and nonsteroidal anti-inflammatory drugs, relieve musculoskeletal pain and reduce inflammation in soft tissues.

anatomy
The study of the structure of the body and its component parts, especially the human body. Anatomy is distinct from physiology, which is the study of bodily functions.

angiosperm
There are over 250,000 different species of angiosperm, or flowering plants, found in a wide range of habitats. A flowering plant is one in which the **seeds** are enclosed within an ovary, which ripens into a **fruit**. Angiosperms are divided into monocotyledons (single seed leaf in the embryo) and dicotyledons (two seed leaves in the embryo). They include the majority of flowers, herbs, grasses, and trees with the exception of conifers.

> **FOSSIL FLOWERS**
>
> Evidence of **fossil** angiosperms has been found dating from the Jurassic era, 208–146 million years ago, and specimens that seem very similar to modern examples have been found from the Cretaceous period, 146–65 million years ago.

Angiosperms are seed plants, differing from **gymnosperms** in that ovules and seeds are protected within a structure called the carpel. **Fertilization** occurs by male gametes, or sex cells, passing into the ovary from a pollen tube. After fertilization the ovule develops into the seed while the ovary wall develops into the fruit.

animal

A member of the kingdom *Animalia*, one of the major categories of living things. Animals are heterotrophs, which means that they obtain their energy from organic substances produced by other organisms. In the past, it was common to include the single-celled **protists** with the animals, but these are now classified separately, together with single-celled plants, as Protista. Thus all animals are multicellular. They have **eukaryotic** cells (the genetic material is contained within a distinct **nucleus**) bounded by a thin **cell** membrane rather than the thick cell wall of plants. Most animals are capable of moving around for at least part of their **life cycle**. The study of animals is **zoology**.

Animals can be divided into three feeding types:
- herbivores eat plants and plant products
- carnivores eat other animals
- omnivores eat both.

Since few animals can digest cellulose, herbivores have either cellulose-digesting bacteria or protozoa in their guts, or grinding mechanisms, such as the large flattened teeth of elephants, to release the plant protoplasm from its cellulose-walled cells. Carnivores are adapted for hunting and eating flesh, with well-developed sense organs and fast reflexes, and weapons such as sharp fangs, claws, and stings. Omnivores eat whatever they can find, and often scavenge among the remains of carnivores' prey; because of the diversity of their diet, they have more versatile teeth and guts than herbivores or carnivores.

> **OLDEST ANIMALS**
>
> The oldest land animals known date back 440 million years. Their remains were found in 1990 in a sandstone deposit near Ludlow, Shropshire, UK, and included fragments of two centipedes a few centimetres long and a primitive spider measuring about l mm.

Many animals are adapted for a parasitic way of life, living on other animals or plants, and feeding solely by absorbing fluids from their hosts. Some animals absorb food directly into their body cells; others have a digestive system in which food is prepared for absorption by body tissues.

anode
The positive **electrode** of an electrolytic **cell**, towards which negative particles (anions), usually in solution, are attracted. See **electrolysis**. A metal is anodized by making it the anode in an electrolytic cell. Oxygen is produced at the anode and combines with the **metal** to form a film of oxide that protects the metal from corrosion.

antenna
An appendage or 'feeler' found on the heads of some **animals**. **Insects**, centipedes, and millipedes each have one pair of antennae but crustaceans, such as shrimps, have two pairs. In insects, the antennae are involved with the senses of smell and touch; they are frequently complex structures with large surface areas that increase the ability to detect scents.

anthropic principle
The idea that the **universe** has to be the way it is because if it were any different we would not be here to observe it. The principle arises from the observation that if the laws of science were even slightly different, it would have been impossible for intelligent life to evolve. For example, if the **electric charge** on the **electron** were only slightly different, stars would have been unable to burn hydrogen and produce the chemical **elements** that make up our bodies.
- *The Strong Anthropic Principle:* the universe must inevitably, at some stage in its history give rise to intelligent life that can observe it.
- *The Weak Anthropic Principle:* there may be other universes where we could not exist, but we would not be able to observe them.

anthropology
The study of humankind. Anthropology is a uniquely Western social science. It investigates the cultural, social, and physical diversity of the human species, both past and present. It is divided into two broad categories: biological or physical anthropology, which attempts to explain human biological variation from an evolutionary perspective; and the larger field of

social or cultural anthropology, which attempts to explain the variety of human cultures. Social anthropology differs from sociology insofar as anthropologists are concerned with cultures and societies other than their own. Anthropology's primary method involves the researcher living for a year or more in another culture, speaking the local language and participating in all aspects of everyday life. Current concerns include ethnohistory, art, migration, ethnological museums, ethnicity, and how different peoples experience and construct time, space, and landscape.

antibiotic

A **drug** derived from living organisms such as **fungi** or **bacteria** that kills or inhibits the growth of other bacteria and fungi. The earliest antibiotics, the penicillins, came into use from 1941 following their accidental discovery in 1928 by Alexander **Fleming** when he observed that a mould growing on a dish containing bacteria had killed the bacteria around it. The mould was similar to that found on stale bread.

Broad-spectrum antibiotics are effective against a wide range of bacteria, while others are specific to a single strain of bacteria. Increasingly, bacteria have become resistant to the antibiotics we use. The emergence of multiple drug resistant bacteria, such as *Staphylococcus*, which resist the most powerful antibiotics we have, is a worrying trend.

> **AGAINST LIFE**
>
> The word antibiotic, first used in 1945, comes from *antibiosis*, against life, which was coined in the 19th century to describe a type of natural competition between microbes.

antibody

One of the defence mechanisms brought into play when foreign or invading substances enter the body. Antibodies are **proteins** produced in the blood by lymphocytes, a type of white blood cell. They react to the presence of **antigens**. Each antibody acts against only one kind of antigen. The antibody combines with the antigen to form a 'complex'. This action may render the antigen harmless, or it may destroy the **micro-organism** by setting off chemical changes that cause them to self-destruct. In other cases, the formation of a 'complex' will cause antigens to form clumps that can then be detected and engulfed by other white blood cells, such as

> **RESISTANCE**
>
> Many **diseases** can only be contracted once because the antibodies produced to deal with them remain in the blood after the infection has passed, preventing any further invasion. Vaccination boosts a person's resistance to disease by triggering the production of antibodies specific to particular infections.

macrophages and phagocytes. Antibody production is only one aspect of **immunity** in **vertebrates**.

See also: *vaccine.*

antigen

Any substance that causes the production of **antibodies** by the body's immune system. Common antigens include the **proteins** carried on the surface of **bacteria**, **viruses**, and pollen grains. The proteins of incompatible **blood groups** or tissues also act as antigens, and this has to be taken into account in medical procedures such as blood transfusions and organ transplants.

See also: *immunity, vaccine.*

antimatter

A form of **matter** that is made up of antiparticles. The attributes of elementary particles, such as **electrical charge**, magnetic moment, and spin are opposite in their equivalent antiparticles. Both particle and antiparticle will have the same **mass**, however. When a particle and its antiparticle collide both are annihilated with a huge release of **energy**. Antiparticles have been created in particle accelerators, such as those at CERN, home of the world's largest particle accelerator, the Large Electron Positron Collider (LEP), in Geneva, Switzerland, and at Fermilab in the USA.

> **FLEETING EXISTENCE**
>
> In 1996 physicists at CERN created the first **atoms** of antimatter: nine atoms of antihydrogen that survived for just 40 nanoseconds (40 billionths of a second).

antioxidant
Any substance that slows the deterioration of fats, oils, paints, plastics, and rubbers by preventing **oxidation**. When used as food **additives**, antioxidants prevent fats and oils from becoming rancid when exposed to air, and thus extend their shelf-life. Vegetable oils contain natural antioxidants, such as vitamin E, which prevent spoilage, but antioxidants are nevertheless added to most oils. They are not always listed on food labels because if a food manufacturer buys an oil to make a food product, and the oil has antioxidant already added, it does not have to be listed on the label of the product.

antiseptic
Any substance that kills or inhibits the growth of disease-causing **micro-organisms** but which is essentially non-toxic to human body cells. The use of antiseptics was pioneered by the surgeon Joseph Lister (1827–1912). Lister was influenced by Louis **Pasteur's** work on micro-organisms and disease. He introduced dressings soaked in carbolic acid and strict rules of hygiene to combat wound sepsis in hospitals. Sepsis was at this time thought to be a kind of combustion caused by exposing moist body tissues to **oxygen**. Under Lister's regime death rates in hospitals due to infection fell dramatically.

Common antiseptics used today include hydrogen peroxide and ethanol, which are used to treat minor wounds.

arachnid
A type of **arthropod** of the class Arachnida, which includes spiders, scorpions, ticks, and mites. They differ from **insects** in possessing only two main body regions, the cephalothorax and the abdomen, and in having eight legs. The cephalothorax has a pair of grasping or piercing appendages, called the chelicerae, and a pair of pedipalps, used for manipulation or as sense organs. The four pairs of legs are also located on the cephalothorax.

Arachnids are generally carnivorous, feeding on the body fluids of their prey or producing **enzymes** to dissolve them.

Archaea
The Archaea are thought to be related to the earliest life forms, which appeared about 4 billion years ago, when there was little oxygen in the Earth's atmosphere. All are strict **anaerobes**, that is, they are killed by oxygen. The archaeans were originally classified as **bacteria**, but in 1996

> **EXTREMOPHILES**
>
> Archaeans are found in undersea vents, hot springs, the salt water of the Dead Sea, and salt pans. Because of the extreme conditions they are found in, the archeans are sometimes called extremophiles.

scientists investigating the genetic make-up of *Methanococcus jannaschii*, an archaean that lives in undersea vents at temperatures around the **boiling point** of water, discovered that 56% of its **genes** were unlike those of any other organism. The archaeans were given the status of their own kingdom in the living world.

Archimedes' principle

The **weight** of the **liquid** displaced by a floating body is equal to the weight of the body. The principle is often stated in the form: 'an object totally or partially submerged in a fluid displaces a **volume** of fluid that weighs the same as the apparent loss in weight of the object (which, in turn, equals the upwards force, or upthrust, experienced by that object – its **buoyancy**).' It was named for the Greek mathematician Archimedes (287–212 BC) who, legend has it, sprang from his bath with a cry of 'Eureka!' when the idea came to him. In fact, he never did state the principle, although it is linked with some of his discoveries.

Aristotle (384–322 BC)

Greek philosopher who advocated reason and moderation. He maintained that sense experience is our only source of knowledge, and that by reasoning

> **ARISTOTLE'S LIFE**
>
> | | |
> |---|---|
> | **384 BC** | Aristotle is born in Stagira in Thrace. |
> | | Studies in Athens at the Academy founded by Plato. |
> | | Opens a school at Assos where he marries Pythias, niece and adopted daughter of Hermeias, ruler of Atarneus. |
> | **c. 344** | Aristotle moves to Mytilene in Lesvo to study natural history. |
> | **342** | He accepts an invitation from Philip II of Macedon to tutor Philip's son Alexander the Great. |
> | **335** | He opens a school in the Lyceum in Athens. |
> | **323** | Aristotle fled to Chalcis when Alexander dies. |
> | **324** | Aristotle dies in Chalis. |

we can discover the essences of things, that is, their distinguishing qualities. In his works on ethics and politics, he suggested that human happiness consists in living in conformity with nature. Of Aristotle's works, around 22 treatises survive, dealing with, among other things, **physics**, **astronomy**, meteorology, **biology**, and psychology. According to Aristotle's laws of motion, bodies moved upwards or downwards in straight lines. Aristotle's work in astronomy included proving that the Earth was spherical. He observed that the Earth cast a circular shadow on the Moon during an eclipse and he pointed out that as one travelled north or south, the stars changed their positions.

arthropod

A member of the Arthropoda, a division of the **animal** kingdom that includes:

- *arachnids* (spiders, scorpions, ticks, and mites)
- *crustaceans* (shrimps, crabs, etc.)
- *myriapods* (centipedes and millipedes)
- *insects*.

Arthropods are **invertebrate** animals with jointed legs and segmented bodies divided into distinct, specialized regions, for example the head, thorax, and abdomen of **insects**. The body has a horny or chitinous outer body called a cuticle that acts as a protective exoskeleton. This is shed periodically and replaced as the animal grows.

> The Arthropoda comprise over 1 million species, the largest group in the animal kingdom.

asexual reproduction

Reproduction without the need for two parents. Asexual reproduction does not involve the manufacture and **fusion** of sex **cells** and every asexual organism can reproduce on its own. This can lead to a rapid **population** build-up. This method carries a clear advantage in that there is no need to search for a mate or to develop complex pollinating mechanisms. The disadvantage of asexual reproduction arises from the fact that only identical individuals, or **clones**, are produced – there is no **variation** and therefore fewer opportunities for **evolution** other than by **mutation**. To offset this disadvantage many asexually reproducing organisms are capable of reproducing sexually at some stage in their **life cycle** as well.

Asexual processes include the parent organism splitting into two or more 'daughter' organisms, and budding, in which a new organism is formed initially as an outgrowth of the parent organism. The asexual reproduction of spores, as in ferns and mosses, is also common and many plants reproduce asexually by means of runners, rhizomes, bulbs, and corms.

astronomy

The science of the celestial bodies: the Sun, the Moon, and the planets; the stars and galaxies; and all other objects in the **universe**. It is concerned with their positions, motions, distances, and physical conditions and with their origins and evolution.

astrophysics

The study of the physical nature of stars, galaxies, and the universe. It began with the development of spectroscopy, the analysis of **spectra**, in the 19th century, which allowed astronomers to analyse the composition of stars from their **light**. Astrophysicists view the universe as a vast natural laboratory in which they can study **matter** under conditions of **temperature**, **pressure**, and **density** that are unattainable on Earth.

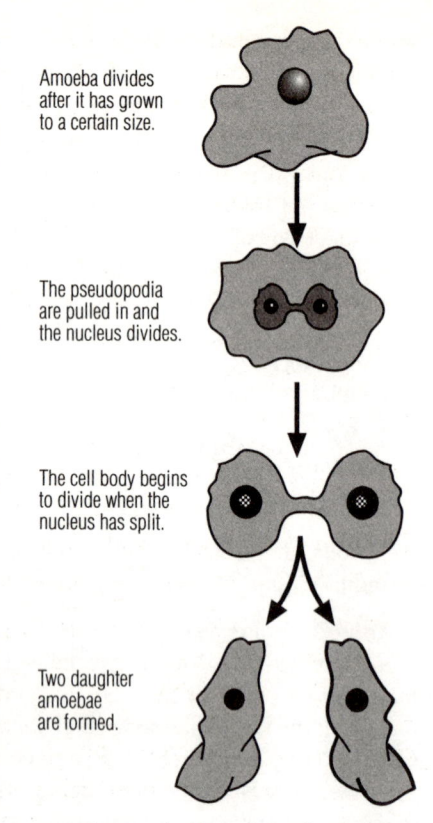

Amoeba divides after it has grown to a certain size.

The pseudopodia are pulled in and the nucleus divides.

The cell body begins to divide when the nucleus has split.

Two daughter amoebae are formed.

asexual reproduction *Asexual reproduction is the simplest form of reproduction, occurring in many simple plants and animals. Binary fission, shown here occurring in an amoeba, is one of a number of asexual reproduction processes.*

atom

The smallest unit of **matter** that can take part in a chemical reaction, and which cannot be broken down chemically into anything simpler. The core of the atom is the **nucleus**, a dense body only one ten-thousandth the diameter of the atom itself. The simplest nucleus, that of hydrogen, comprises a single stable positively charged particle, the **proton**. Nuclei of other elements contain more protons and additional particles, called **neutrons**, of about the same **mass** as the proton but with no **electrical charge**.

> **LARGEST OF THE SMALL**
>
> The largest atom, that of caesium, has a diameter of 0.0000005 mm/ 0.00000002 in.

- Negatively charged **electrons** are arranged around the nucleus of an atom in distinct energy levels called orbitals or shells, each of which can contain a certain maximum number of electrons.
- The atomic number of an **element** indicates the number of electrons in a neutral atom, the negative charges on the electrons balancing the positive charges on the protons.
- The outermost shell is known as the **valency** shell and contains the electrons involved in reactions with other atoms. The chemical properties of an element are determined by the ease with which its atoms can gain or lose electrons from its valency shell.
- In **ions**, the electron shells contain more or fewer electrons than are required for a neutral atom, generating negative or positive charges.

See also: *fundamental forces, isotopes, subatomic particles.*

atomic clock

A timekeeping device regulated by processes occurring in **atoms** and **molecules**, such as atomic vibration.

- *Ammonia clock*, invented at the US National Bureau of Standards in 1948, was regulated by measuring the speed at which the nitrogen atom in an ammonia molecule vibrated back and forth. The rate of molecular vibration is not affected by **temperature**, **pressure**, or other external influences, making it a very accurate timekeeper.
- *Caesium clock* is even more accurate. Because of its internal structure, a caesium atom produces or absorbs radiation of a very precise frequency (9,192,631,770 Hz) that varies by less than one part in 10 billion. This frequency has been used to define the second, and is the basis of atomic clocks used in international timekeeping.
- *Hydrogen maser clocks*, based on the radiation from hydrogen atoms, are the most accurate of all. The hydrogen maser clock at the US Naval Research Laboratory, Washington, DC, is estimated to lose one second in 1,700,000 years. Cooled hydrogen maser clocks could theoretically be accurate to within one second in 300 million years.

CLOCK ADJUSTMENTS

Atomic clocks are so accurate that adjustments must be made periodically to bring atomic clock time and calendar time into line. In 1997 the northern hemisphere's summer was longer than usual – by one second. The extra second was added to the world's time at precisely 23 hours, 59 minutes, and 60 seconds on 30 June 1997. The adjustment was called for by the International Earth Rotation Service in Paris, which monitors the difference between Earth time and atomic time.

ATP (or adenosine triphosphate)

An abbreviation for a nucleotide **molecule** found in all cells. It can yield large amounts of energy, and is used to drive the thousands of biological processes needed to sustain life, growth, movement, and reproduction. Green plants use light energy to manufacture ATP as part of the process of **photosynthesis**. In animals, ATP is formed by the breakdown of glucose molecules, usually obtained from the carbohydrate component of a diet, in the process called **respiration**. ATP is the driving force behind muscle contraction and the synthesis of complex molecules needed by individual cells.

autotroph

Any living organism that can obtain all the food it needs by making organic substances from inorganic ones by using light or chemical energy. Autotrophs are the primary producers in all **food chains** since the materials they synthesize and store are the energy sources of all other organisms. All green plants and many planktonic organisms are autotrophs, using sunlight to convert carbon dioxide and water into sugars by **photosynthesis**. Some **bacteria** use the chemical energy of sulphur **compounds** to synthesize organic substances. The total **biomass** of autotrophs is far greater than that of **animals**, reflecting the dependence of animals on plants, and the ultimate dependence of almost all life on energy from the Sun.

Avogadro, Amedeo, Conte di Quaregna (1776–1856)

Italian physicist, one of the founders of physical **chemistry**. Avogadro was born in Turin, where he spent his whole academic career. Avogadro made it clear that **gas** particles need not be individual **atoms** but might consist of **molecules**, the term he introduced to describe combinations of atoms. No previous scientist had made this fundamental distinction between the atoms of a substance and its molecules. The **electrolysis** of water (to form hydrogen and oxygen) produces twice as much hydrogen (by volume) as oxygen. Avogadro reasoned that each molecule of water must contain

> **AVOGADRO'S HYPOTHESIS**
>
> Equal volumes of all gases, when at the same **temperature** and **pressure**, have the same numbers of molecules.

hydrogen and oxygen atoms in the proportion of 2 to 1. Also, because the oxygen gas collected weighs eight times as much as the hydrogen, oxygen atoms must be 16 times as heavy as hydrogen atoms.

Avogadro's number
Avogadro's number or constant is equal to 6.022045×10^{23} – the number of atoms in 12 grams of **carbon**. Twelve is also the atomic mass number of carbon. A weight in grams of any **element** that is the same as the atomic mass number of that element will contain Avogadro's number of atoms.

B

background radiation
Radiation that is always present in the environment. By far the greater proportion (87%) of it is emitted from natural sources such as the traces of radioactive minerals that occur naturally in the environment and even in the human body, and by radioactive gases such as radon and thoron, which are found in soil and may seep upwards into buildings. **Radiation** from space also contributes to the background level.

bacteria (singular bacterium)
A microscopic single-celled **prokaryote** organism that is a member of the kingdom Bacteria, one of the major divisions of life. Bacteria are widespread, and are found in soil, air, and water, and as **parasites** on and in

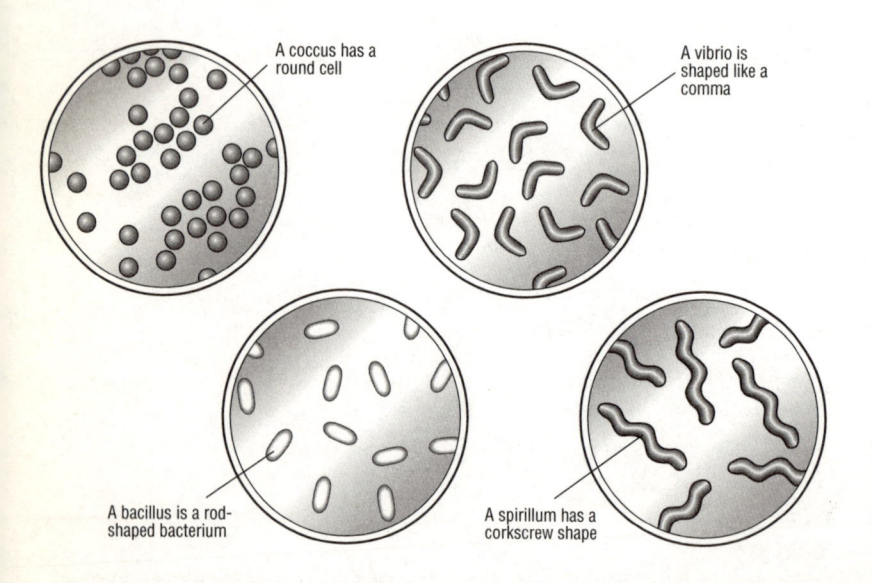

bacteria *Four different types of bacterial cells: cocci, bacilli, vibrios, and spirilla.*

other living things. Bacteria usually reproduce by binary fission (dividing into two equal parts), and this may occur approximately every 20 minutes under ideal conditions. Bacterial **cells** do not have a central **nucleus** containing their genetic material. They have a large loop of **DNA**, sometimes called a bacterial **chromosome**. In addition there are often small, circular pieces of DNA known as plasmids that also carry genetic information. These plasmids can move readily from one bacterium to another, even though the bacteria may be of different **species**. Some plasmids confer **antibiotic** resistance on the bacteria they inhabit. Bacteriologists believe that around 3 million species may actually exist although only some 4,000 have been classified as yet.

> **OUTNUMBERED!**
>
> There are ten times more bacterial cells than human cells in and on the human body.

Bacterial classifications
- *cocci* are round or oval
- *bacilli* are rodlike
- *spirilla* are spiral
- *vibrios* are shaped like commas.

Bacteria can also be classified into Gram positive or Gram negative according to their reactions to certain stains, or dyes, used in microscopy.

Beneficial bacteria
Bacteria can be used to:.
- break down waste products
- make butter, cheese, and yoghurt
- cure tobacco
- tan leather
- extract minerals from mines
- digest some pollutants.

battery
An energy-storage device that allows the release of **electricity** on demand. It is made up of one or more electrical **cells**. Primary-cell batteries are dis-

posable; secondary-cell batteries, or accumulators, are rechargeable. Primary-cell batteries are an extremely uneconomical form of energy, since they produce only 2% of the power used in their manufacture.

The common dry cell is a primary-cell battery and consists of a central **carbon** electrode immersed in a paste of manganese dioxide and ammonium chloride as the electrolyte. The zinc casing forms the other **electrode**. The lead–acid car battery is a secondary-cell battery. The car's generator continually recharges the battery when the engine is running. It consists of sets of lead (positive) and lead peroxide (negative) plates in an electrolyte of sulphuric **acid** (battery acid).

PLASTIC BATTERY

The first all-plastic battery was created in the USA in July 1996. In 1997, a credit-card sized version of the plastic battery was introduced by its US inventors in Baltimore, Maryland. It produces 2.5 volts of electricity and does not have the **toxins** that can leak from ordinary batteries.

Becquerel, (Antoine) Henri (1852–1908)

French physicist who was the first to discover **radioactivity**. He shared a Nobel prize with Marie and Pierre **Curie** in 1903.

Becquerel was born and educated in Paris. In 1875 he began private scientific research, investigating the behaviour of polarized **light** in magnetic fields and in **crystals**. The discovery of **X-rays** in 1896 prompted Becquerel to investigate fluorescent crystals for the emission of X-rays, and in so doing he accidentally discovered radioactivity in uranium salts in the same year. Becquerel subsequently investigated the radioactivity of radium, and showed in 1900 that it consists of a stream of **electrons**. In the same year, Becquerel also obtained evidence that radioactivity results from the transformation of one **element** into another.

Big Bang

The event that marked the birth of the known **universe**. At the time of the Big Bang, the entire universe existed as a **singularity** – an infinitely small, infinitely dense point where none of the known laws of **space and time** apply. For some unknown reason this point suddenly expanded in a huge outpouring of space and **energy** some 10 to 20 billion years ago. As the universe expanded it cooled and all the matter in it condensed from the energy, like water droplets forming from steam. According to inflation theory, the universe underwent a super rapid period of expansion shortly after the Big Bang when space was expanding faster than the **speed of light**. This accounts for its current large size and uniform nature of the observable universe. The inflationary theory is supported by the most recent observations of the cosmic **background radiation**.

> **BIG BANG**
>
> Scientists have calculated that 10^{-36} seconds (one million-million-million-million-million-millionth of a second) after the Big Bang, the universe was the size of a pea, and the temperature was 10 billion million million million °C (18 billion million million million °F). One second after the **Big Bang**, the temperature was about 10 billion °C (18 billion°F).

biochemistry

The branch of science that is concerned with the chemical activities that go on in living organisms. Its study has led to an increased understanding of life processes, such as those by which organisms synthesize essential chemicals from food materials, store and generate energy, and pass on their characteristics through their genetic material. A great deal of medical research is concerned with the ways in which these processes are disrupted. Biochemistry also has applications in agriculture and in the food industry (for instance, in the use of **enzymes**).

biodegradable

Materials that are capable of being broken down by living organisms, principally **bacteria** and **fungi**, are said to be biodegradable. Natural processes of decay lead to compaction and liquefaction in naturally biodegradable substances, such as food and sewage. This releases nutrients that can be taken up by plants and recycled through the **ecosystem**. Where it takes place on a large scale, in rubbish tips for example, this process can have some disadvantageous side effects, such as the release of methane, an explosive greenhouse gas. However, the technology now exists for waste tips to collect methane in underground pipes, drawing it off and using it as a cheap source of energy. Biogas digesters, which produce methane gas from animal waste, are used as a cheap source of energy in many developing countries.
See also: *biofuel.*

biodiversity
A measure of the variety of life on Earth. Estimates of the number of **species** living on the planet vary widely because many species-rich **ecosystems**, such as tropical forests, contain unexplored and unstudied **habitats**.

The maintenance of biodiversity is important for ecological stability and as a resource for research into, for example, new **drugs** and crops. The most significant threat to biodiversity comes from the destruction of rainforests and other habitats in the southern hemisphere. The destruction of habitats in the 20th century is believed to have resulted in the most severe and rapid loss of biodiversity in the history of the planet.

> The majority of small organisms remain unknown to science and even larger species, such as **mammals**, still come to light from time to time.

biofuel
Any fuel produced from organic matter, either directly from **plants** or indirectly from industrial, commercial, domestic, or agricultural wastes. There are three main methods for the development of biofuels: the burning of dry organic wastes (such as household refuse, industrial and agricultural wastes, straw, wood, and peat); the **fermentation** of wet wastes (such as animal dung) in the absence of oxygen to produce biogas (containing up to 60% methane), or the fermentation of sugar cane or corn to produce **alcohol** and esters; and energy forestry (producing fast-growing wood for fuel).

biological clock
A mechanism that seemingly exists within many living things that produces regular periodic changes in activity, such as the start of a breeding season or a period of **hibernation**. Such clocks are known to exist in almost all animals, and also in many plants, fungi, and unicellular organisms. The first biological clock **gene** in plants was isolated in 1995 by a team of researchers in the USA. In higher organisms, there appears to be a series of clocks of graded importance. For example, although body temperature and activity cycles in human beings are normally 'set' to 24 hours, the two cycles may vary independently, showing that two clock mechanisms are involved.

biological control
The control of pests such as **insects** and **fungi** through biological means, rather than the use of chemicals.

Biocontrol methods
- breeding resistant crop strains
- inducing sterility in the pest
- infecting the pest species with disease organisms
- introducing a natural predator.

Biological control tends to be naturally self-regulating, but as **ecosystems** are so complex, it is difficult to predict all the consequences of introducing a biological controlling agent. Ladybirds are sometimes used to control aphids as both adults and **larvae** feed on them. In 1998, French researchers patented a method of selectively breeding hardy flightless ladybirds for use in biological control, as captive **populations** are far more effective than mobile ones.

> **PERILS OF BIOCONTROL**
>
> The cane toad was introduced to Australia in the 1940s to eradicate a beetle that was destroying sugar beet. However, the poisonous cane toad has few Australian predators and has subsequently become a pest itself, spreading throughout eastern and northern Australia.

biology

Biology is the science of life. It includes all the life sciences – for example, **anatomy** and physiology (the study of the structure and function of living things), cytology (the study of **cells**), **zoology** (the study of **animals**), **botany** (the study of **plants**), **ecology** (the study of **habitats** and the interaction of living **species**), animal behaviour, embryology, and taxonomy. Increasingly biologists have concentrated on molecular structures to study **biochemistry**, biophysics, and **genetics** (the study of inheritance and **variation**).

bioluminescence

The production of **light** by living organisms. Many deep-sea fishes, crustaceans, and other marine animals are able to produce light and on land bioluminescence is seen in some nocturnal insects such as glow-worms and fireflies, and in certain **bacteria** and **fungi**. Light is usually produced by the **oxidation** of the chemical luciferin, a **reaction** catalysed by the **enzyme** luciferase. This reaction is unique, being the only known biological oxidation that does not produce heat. Animal luminescence is involved in communication, **camouflage**, or the luring of prey, but its function in other organisms is unclear.

biomass
The total mass of living organisms present in a given area. It may be specified for a particular species (such as earthworm biomass) or for a general category (such as **herbivore** biomass). Measurements of biomass can be used to study interactions between organisms, the stability of those interactions, and variations in **population** numbers.

bird
There are nearly 8,500 **species** of birds, all members of the class Aves, the biggest group of land **vertebrates**.

Bird characteristics
- feathers
- warm blood
- wings
- breathing through lungs
- egg-laying by the female
- bipedal; feet usually adapted for perching and never having more than four toes
- hearing and eyesight well developed
- sense of smell usually poor
- no teeth.

Most birds can fly, but some groups (such as ostriches) are flightless, and others include flightless members. Many communicate by sounds (nearly half of all known species are songbirds) or by visual displays, many species are brightly coloured, usually the males. Birds have highly developed patterns of instinctive behaviour.

> **THREATENED BIRDS**
>
> According to the Red List of endangered species published by the World Conservation Union (IUCN) for 1996, 11% of bird species are threatened with **extinction**.

Territorial behaviour
Within individual species, males may defend an area or territory against competing males. The size of territory varies between species. The gannet, a seabird which nests in dense colonies on cliffs and rocky islets, may defend an area encompassing only the extent to which the sitting female can jab with her bill, whereas the robin defends a territory of half a hectare.

Once a songbird has selected its territory, it will sing to fulfil the dual purpose of advertising its presence to rival territory holders and attracting a female. The song must signify to the female that the singer is a territory-holding male of the correct species in breeding condition, and is therefore important as a species-isolating mechanism. Closely related species that overlap in some habitats, for example the chiffchaff and willow warbler, often have conspicuously different songs. A song must be sufficiently stereotyped to be recognizable to other members of the species, but there is still room for much variation within these confines.

Different bird species occupying the same area usually have different food requirements, different nesting habits, and specific song and courtship behaviour, which not only prevents interbreeding but also reduces competition between species.

black hole

A point in space where **gravity** has caused **matter** to collapse and become incredibly dense. The gravitational force of a black hole is so great that nothing – not even **light** – can escape it. Black holes are thought to form when massive stars run out of fuel at the end of their lives and nuclear forces within the star are no longer enough to prevent their collapse due to gravitational forces. Obviously, no one has seen a black hole, but their presence can be detected by observing the **energy** emitted by material that is swept into the hole.

BIG BLACK HOLES

The Hubble Space Telescope discovered evidence of a black hole 300 million times the **mass** of the Sun in 1997. It is located in the middle of galaxy M84 about 50 million light years from Earth. Similar sized holes may exist at the heart of other galaxies, including our own.

blood

The fluid circulating in the arteries, veins, and capillaries of **vertebrate** animals, and also in those **invertebrates** that possess a closed circulatory system. Blood carries nutrients and oxygen to each body cell and removes waste products, such as carbon dioxide. It is also

Blood cells constantly wear out and die and are replaced from the bone marrow. Red blood cells die at the rate of 200 billion per day but the body produces new cells at an average rate of 9,000 million per hour.

important in the immune response and, in many animals, in the distribution of heat throughout the body.

In humans blood makes up 5% of the body weight, occupying a volume of 5.5 l/10 pt in the average adult. It is composed of a fluid called plasma, in which are suspended microscopic cells of three main varieties: red cells which transport oxygen around the body, white cells which protect the body from disease by ingesting invading bacteria and producing **antibodies**, and platelets which assist in the clotting of blood.

blood group

Blood is classified according to the presence or otherwise of certain **antigens** on the surface of its red cells. Red blood cells of one individual may carry **molecules** on their surface that act as antigens in another individual whose red blood cells lack these molecules. The two main antigens are designated A and B. These give rise to four blood groups: having A only (A), having B only (B), having both (AB), and having neither (O). Each of these groups may or may not contain the rhesus factor. Correct typing of blood groups is vital in transfusion, since incompatible types of donor and recipient blood will result in coagulation, with possible death of the recipient.

Bohr, Niels Henrik David (1885–1962)

Danish physicist whose theoretical work did much to establish the structure of the **atom** and the validity of **quantum theory**. For this work he was awarded the Nobel Prize for Physics in 1922. He explained the structure and behaviour of the **nucleus**, as well as the process of nuclear **fission**.

In 1913, Bohr developed his theory of atomic structure by applying quantum theory to the observations of **radiation** emitted by atoms. Ten years earlier, Max **Planck** had proposed that radiation is emitted or absorbed by atoms in discrete units, or quanta, of **energy**. Bohr postulated that an atom may exist in only a certain number of stable states, each with a certain amount of energy in which **electrons** orbit the nucleus without emitting or absorbing energy. He proposed that emission or absorption of

---

❝ How wonderful that we have met with a paradox. Now we have hope of making progress. ❞

Niels Bohr, quoted in A Pais *Niels Bohr's Times*, 1991

energy occurs only with a transition from one stable state to another. When a transition occurs, an electron moving to a higher orbit absorbs energy and an electron moving to a lower orbit emits energy. In so doing, a set number of quanta of energy are emitted or absorbed at a particular **frequency**.

boiling point

The temperature at which the application of **heat** converts a **liquid** into **vapour**. The boiling point of water under normal **pressure** is 100°C/212°F. Raising or lowering the pressure will raise or lower the boiling point.

bond

The forces of attraction that hold together **atoms** of one or more types of **element** to form a **molecule**. The principal types of bonding are ionic, covalent, metallic, and intermolecular (such as hydrogen bonding). The type of bond formed depends on the elements concerned and their electronic structure.

- *Ionic or electrovalent bond*: the combining atoms gain or lose **electrons** to become **ions**; for example, in sodium chloride (NaCl) sodium (Na) loses an electron to form a sodium ion (Na$^+$) while chlorine (Cl) gains an electron to form a chloride ion (Cl$^-$); electrical forces hold the oppositely charged ions together.

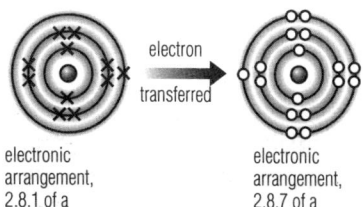

electronic arrangement, 2.8.1 of a sodium atom

electron transferred

electronic arrangement, 2.8.7 of a chlorine atom

becomes a sodium ion, Na$^+$, with an electron arrangement 2.8

becomes a chloride ion, Cl$^-$, with an electron arrangement 2.8.8

bond *The formation of an ionic bond between a sodium atom and a chlorine atom to form a molecule of sodium chloride. The sodium atom transfers an electron from its outer electron shell (becoming the positive ion Na$^+$) to the chlorine atom (which becomes the negative chloride ion Cl$^-$. The opposite charges mean that the ions are strongly attracted to each other. The formation of the bond means that each atom becomes more stable, having a full quota of electrons in its outer shell.*

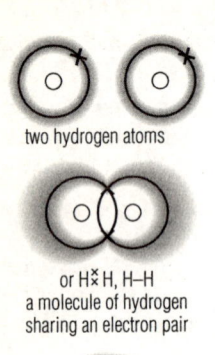

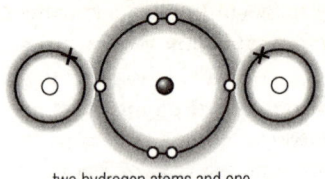

bond *The formation of a covalent bond between two hydrogen atoms to form a hydrogen molecule (H_2), and between two hydrogen atoms and an oxygen atom to form a molecule of water (H_2O). The sharing means that each atom has a more stable arrangement of electrons (its outer electron shells are full).*

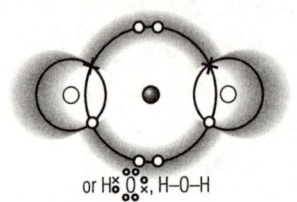

- *Covalent bond*: the atomic orbitals of two atoms overlap to form a molecular orbital containing two electrons, which are thus effectively shared between the two atoms. Covalent bonds are common in organic compounds.
- *Metallic bond*: joins metals in a crystal lattice; valence electrons are shared between all the ions in an 'electron gas'.
- *Hydrogen bond*: a hydrogen atom joined to an electronegative atom, such as nitrogen or oxygen, becomes partially positively charged, and is weakly attracted to another electronegative atom on a neighbouring molecule.

bone

The hard connective tissue that makes up the skeleton of most **vertebrate** animals. Bone is composed of a network of protein fibres, called collagen, impregnated with mineral salts (largely calcium phosphate and calcium carbonate), a combination that gives it great density and strength, comparable in some cases with that of reinforced concrete. Enclosed within this solid matrix are bone cells, blood vessels, and nerves. The interior of the long bones of the limbs consists of a spongy matrix filled with a soft marrow that produces blood cells.

botany

The study of **plants**, including their **classification**, form, function and interaction with the **environment**.

Botanical branches

- taxonomy: the identification and classification of plants
- morphology: the external structure of plants
- plant anatomy: the internal structure of plants
- plant histology: the microscopic structure of plants
- plant physiology: the functioning and life history of plants
- plant **ecology**: distribution and relationships of plants over the Earth's surface
- horticulture
- agriculture
- forestry.

C

camouflage
Colours or structures that allow an animal to blend with its surroundings to avoid detection by other animals. Camouflage can take the form of matching the background colour, of countershading (darker on top, lighter below, to counteract natural shadows), or of irregular patterns that break up the outline of the animal's body. More elaborate camouflage involves closely resembling a feature of the natural environment, as with the stick insect; this is closely akin to **mimicry**. Camouflage is also important as a military technique, disguising either equipment, troops, or a position in order to conceal them from an enemy.

cancer
There are more than 100 types of cancer. This group of **diseases** is characterized by the abnormal proliferation of **cells**. Some cancers, like lung or bowel cancer, are common; others are rare. The likely causes of many remain unexplained.

Triggering agents (carcinogens)
- chemicals such as those found in cigarette smoke
- other forms of smoke
- asbestos dust
- exhaust fumes
- many industrial chemicals
- **X-rays** and other forms of radioactivity.

Some **viruses** carry **genes**, called **oncogenes** that can also trigger the cancerous growth of cells Malignant cancer cells are usually capable only of reproducing themselves (tumour formation) and carry out no other purpose within the body. Malignant cells tend to spread from their site of origin by travelling through the bloodstream or lymphatic system.

Cancer kills about 6 million people a year worldwide.

carbohydrate

Carbohydrates are organic chemical compounds composed of **carbon**, **hydrogen**, and **oxygen** and are produced by green **plants** in **photosynthesis**. They are an important part of a balanced human diet, providing **energy** for life processes including growth and movement.

The simplest carbohydrates are simple sugars (monosaccharides), such as glucose and fructose, and disaccharides, such as fructose and lactose, which are soluble **compounds**, some with a sweet taste. When these basic sugar units are joined together in long chains or branching structures they form polysaccharides, such as starch and glycogen, which often serve as food stores in living organisms.

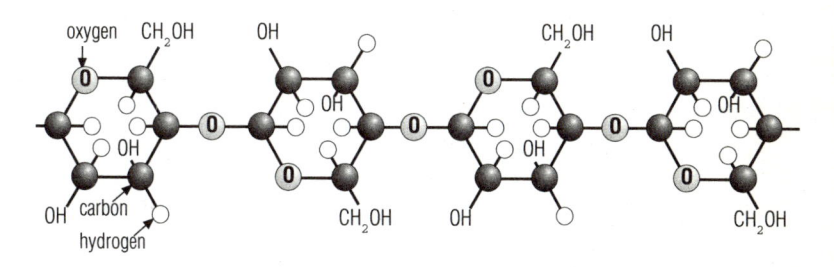

carbohydrate *A molecule of the polysaccharide glycogen (animal starch) is formed from linked glucose ($C_6H_{12}O_6$) molecules. A typical glycogen molecule has 100–1,000 glucose units.*

Even more complex carbohydrates are known, including chitin, which is found in the **cell** walls of **fungi** and the hard outer **skeletons** of **insects**, and cellulose, which makes up the cell walls of plants. Excess carbohydrate intake can be converted into **fat** and stored in the body.

carbon

Carbon is unusual among the elements because its atoms can link with one another in rings or chains, giving rise to innumerable complex **compounds** in carbonaceous rocks such as chalk and limestone; as **hydrocarbons** in petroleum, coal, and natural **gas**; and as a constituent of all organic substances. Of the inorganic carbon compounds, the chief ones are carbon dioxide (CO_2) and carbon monoxide (CO). Carbon is also found on its own as diamond, graphite, and as fullerenes.

Uses of carbon

Carbon is used as a fuel in the form of coal or coke. When added to steel, carbon forms a wide range of alloys with useful properties. In pure form, it is used as a moderator in nuclear reactors; as colloidal graphite it is a good **lubricant**.

carbon cycle

The movement of carbon through the natural world. Carbon dioxide is released into the atmosphere by

> **BUCKYBALLS**
>
> Analysis of interstellar dust led to the discovery of pure carbon molecules, each containing 60 carbon atoms, in a shape similar to that of the geodesic domes designed by US architect and engineer Buckminster Fuller. They were named buckminsterfullerenes (often shortened to 'buckyballs') in his honour.

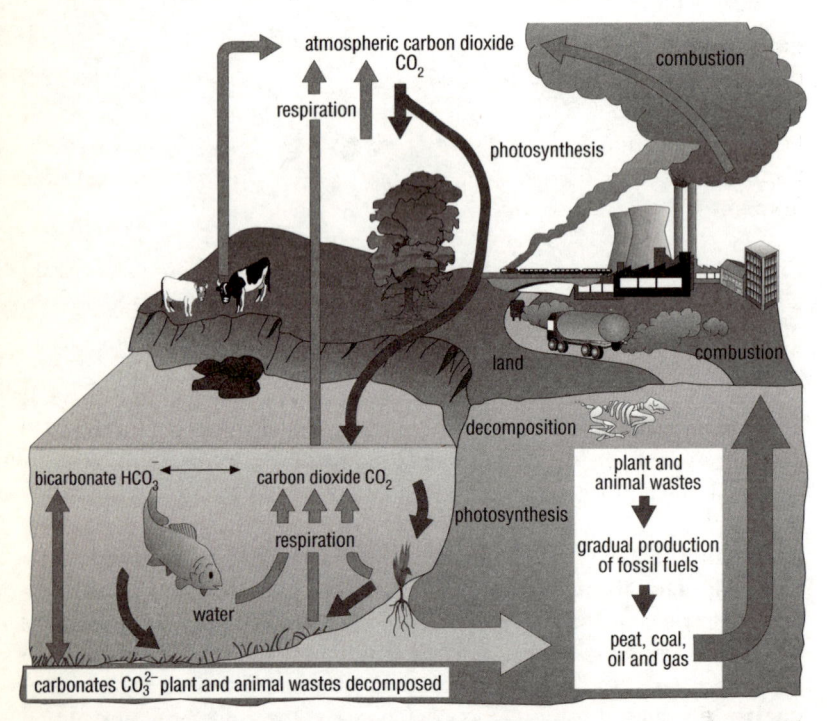

carbon cycle *The carbon cycle is necessary for the continuation of life. Since there is only a limited amount of carbon in the Earth and its atmosphere, carbon must be continuously recycled if life is to continue. Other chemicals necessary for life – nitrogen, sulphur, and phosphorus, for example – also circulate in natural cycles.*

living things as a result of **respiration**. The CO_2 is taken up and converted into **carbohydrates** during **photosynthesis** by **plants** and by organisms such as diatoms and dinoflagellates in the oceanic **plankton** (the **oxygen** is released into the atmosphere as a by-product). When an **animal** eats a plant carbon is transferred from the plant to the animal. Carbon is also released through the **decomposition** of decaying plant and animal matter, and the burning of **fossil** fuels such as coal (fossilized plants).

carcinogen
Any agent that increases the chance of a **cell** becoming cancerous (see **cancer**), including various chemical compounds, some **viruses**, **X-rays**, and other forms of ionizing **radiation**. The term is often used more narrowly to mean chemical carcinogens only.

carnivore
Carnivores are **animals** that eat other animals. The name is especially used to describe members of the order Carnivora, which is made up of mainly flesh-eating **mammals**, including cats, dogs, bears, badgers, and weasels. Carnivores have the greatest range of body size of any mammalian order, from the 100 g/3.5 oz weasel to the 800 kg/1,764 lb polar bear.

Carnivora characteristics
- sharp teeth, small incisors
- a well-developed brain
- a simple stomach
- generally sharp and powerful claws.

CARNIVORE HERBIVORES
Not all carnivores eat meat. Pandas eat bamboo, and civet cats eat fruit.

catalyst
A substance that alters the speed of, or makes possible, a chemical or biochemical **reaction** but remains unchanged at the end of the reaction. In practice most catalysts are used to speed up reactions. **Enzymes** are natural biochemical catalysts.

cathode
The negative **electrode** of an electrolytic **cell**, towards which positive particles (cations), usually in **solution**, are attracted.
See also: *electrolysis.*

cell (biology)

The basic unit of **life**. The cell is the smallest living thing capable of independent existence that can reproduce itself exactly. All living organisms – with the exception of **viruses** – are composed of one or more cells. Single cell organisms such as **bacteria**, protists, and other **micro-organisms** are termed unicellular (single-celled), while **plants** and **animals** which contain many cells are termed multicellular organisms. Each cell has a surrounding membrane, which is a thin layer of **protein** and **fat** that restricts the flow of substances in and out of the cell. In general, plant cells differ from animal cells in that the cell membrane is surrounded by a cell wall made of **cellulose**. Highly complex organisms such as human beings consist of billions of cells, all of which are adapted to carry out specific functions – for instance, groups of these specialized cells are organized into tissues and organs.

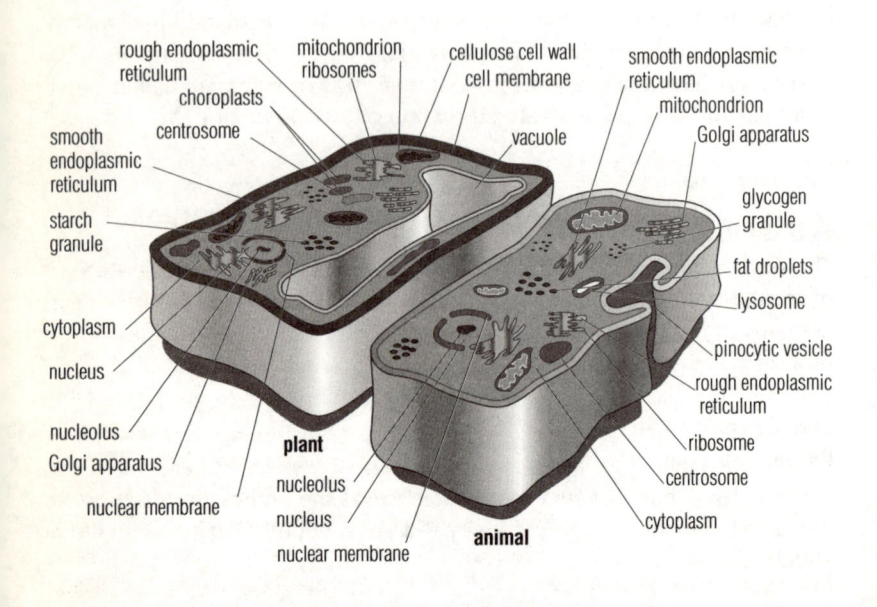

cell *Typical plant and animal cell. Plant and animal cells share many structures, such as ribosomes, mitochondria, and chromosomes, but they also have notable differences: plant cells have chloroplasts, a large vacuole, and a cellulose cell wall. Animal cells do not have a rigid cell wall but have an outside cell membrane only.*

Organelles

Cells contain structures called organelles that perform a variety of tasks.

- *Chloroplasts* (found only in plant cells) contain **chlorophyll** and capture the Sun's energy through **photosynthesis**.
- *Endoplasmic reticulum (ER)* is a system of membranes where fats and proteins are produced.
- The *Golgi apparatus* stores and later transports the proteins manufactured by the ER.
- *Mitochondria* are involved in the cell's metabolic activities that enable the release of **energy**.
- *Ribosomes* carry out protein synthesis.

See also: *cell theory, eukaryote, prokaryote.*

YOU'RE JUST NOT THE SAME PERSON!

Within the human body, about 3 billion cells die every minute; at the same time, these are replaced by about the same number of new cells. External cells, such as skin cells, flake off, while the dead cells from internal organs are passed out of the body with waste products.

---

cell division

The process by which a **cell** divides to form two new cells. When a cell divides its genetic material (**DNA**) is duplicated and each of the new cells contains a complete copy of the original cell's genetic information. This process is called mitosis.

In multicellular organisms, sexual reproduction requires the production of male and female germ cells (sperm and eggs), which contain only half of the normal compliment of DNA. During this process, called meiosis, the cell divides twice, but its **chromosomes** are duplicated only once. Thus, four germ cells are produced, each containing half the normal number of chromosomes. In the male organism the germ cells develop into sperm; in the female they develop into eggs. When a sperm and an egg unite (**fertilization**) a new cell is formed, called a zygote, which has a complete set of chromosomes, and which has received half its genetic information from each parent, thus producing a new individual.

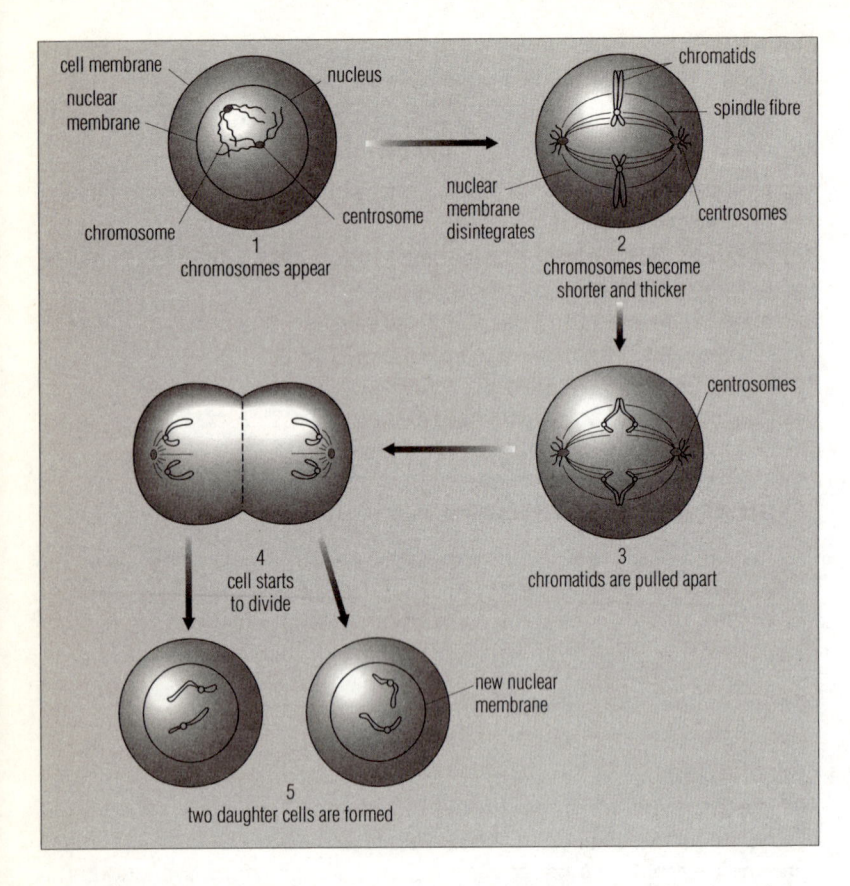

cell division *The stages of mitosis, the process of cell division that takes place when a plant or animal cell divides for growth or repair. The two daughter cells each receive the same number of chromosomes as were in the original cell.*

CELL, ELECTRICAL · 43

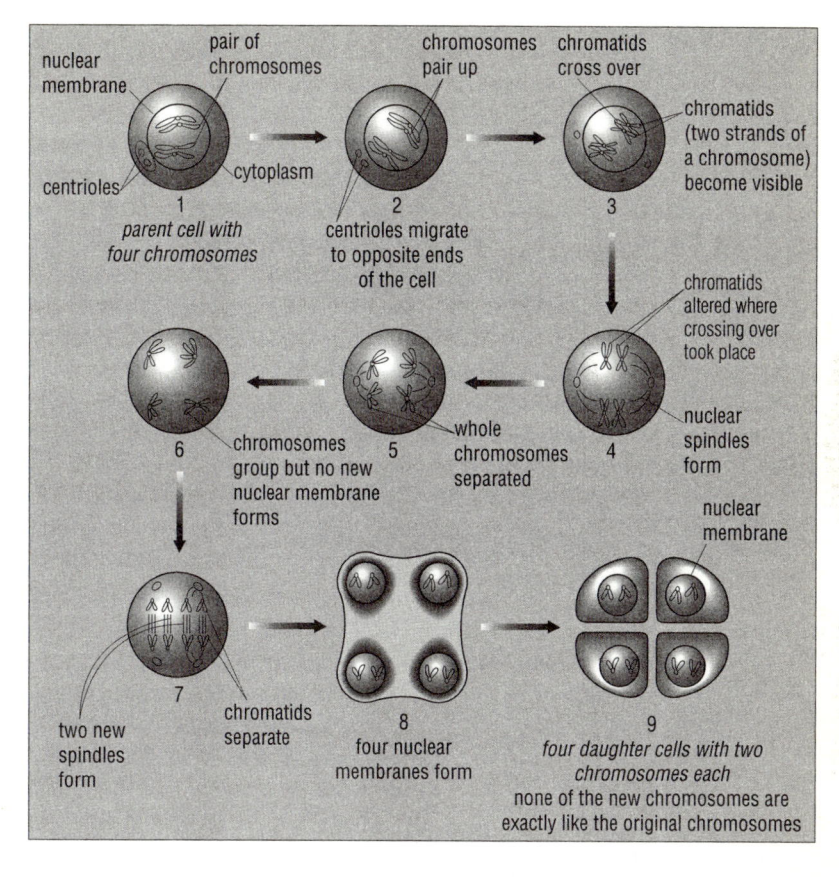

cell division *Meiosis is a type of cell division that produces gametes (sex cells, sperm and egg). This sequence shows an animal cell but only four chromosomes are present in the parent cell (1). There are two stages in the division process. In the first stage (2–6), the chromosomes come together in pairs and exchange genetic material. This is called crossing over. In the second stage (7–9), the cell divides to produce four gamete cells, each with only one copy of each chromosome from the parent cell.*

cell, electrical

A device in which chemical energy is converted into electrical energy. The popular name for an electrical cell is a **battery**, but this actually refers to a

FIRST CELL

The first cell was made by Italian physicist Alessandro Volta in 1800, hence the alternative name voltaic cell.

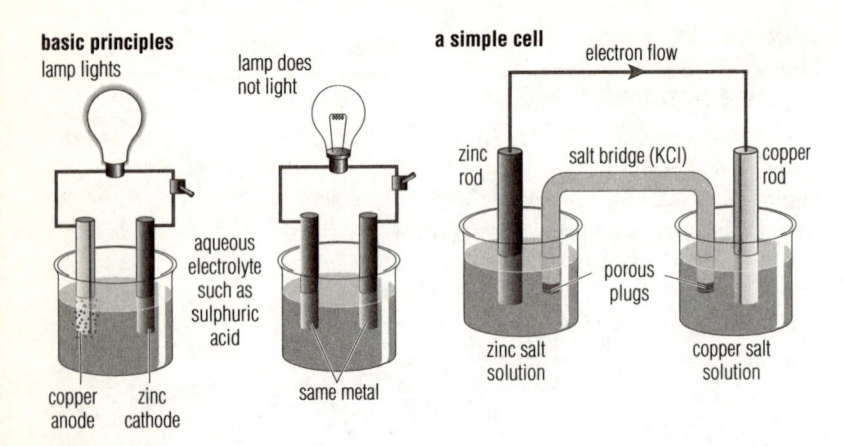

basic principles — lamp lights / lamp does not light / aqueous electrolyte such as sulphuric acid / copper anode / zinc cathode / same metal

a simple cell — electron flow / zinc rod / salt bridge (KCl) / copper rod / porous plugs / zinc salt solution / copper salt solution

cell *When electrical energy is produced from chemical energy using two metals acting as electrodes in a aqueous solution, it is sometimes known as a galvanic cell or voltaic cell. Here the two metals copper (+) and zinc (–) are immersed in dilute sulphuric acid, which acts as an electrolyte. If a light bulb is connected between the two, an electric current will flow with bubbles of gas being deposited on the electrodes in a process known as polarization.*

collection of cells in one unit. Each cell contains two conducting **electrodes** immersed in an electrolyte, in a container. A spontaneous chemical **reaction** within the cell generates a negative charge (an excess of **electrons**) on one electrode, and a positive charge (deficiency of electrons) on the other. The accumulation of these equal but opposite charges prevents the reaction from continuing unless an outer connection (external circuit) is made between the electrodes, allowing the charges to dissipate. When this occurs, electrons escape from the cell's negative terminal and are replaced at the positive, causing a current to flow.

cell theory

English scientist Robert Hooke (1635–1705), first used the word 'cell' in 1665 to describe the structure of plant tissue. Soon afterwards, the Dutch microscopist Anton van **Leeuwenhoek** was able to describe these cells in greater detail. In 1839, with the help of improved microscopes, two German scientists, Matthias Schleiden and Theodor Schwann, formulated the current theory which states that cells are the basic units of construction of all living things. This concept was taken further by German pathologist Rudolf Virchow, who advanced the idea that new cells are formed by the division of existing cells, and that this gives rise to growth and reproduction in plants and animals.

cellulose

Cellulose is the most abundant substance found in the natural world. A complex **carbohydrate** composed of long chains of glucose units it is the principal constituent of the **cell** wall of higher **plants**, giving it rigidity. Cellulose is a vital ingredient in the diet of many **herbivores**. However, no **mammal** produces the **enzyme**, cellulase, necessary for digesting cellulose, instead they must rely on **bacteria** in their guts that do possess the enzyme.

Uses of cellulose
- rope-making
- textiles (linen, cotton, viscose, and acetate)
- plastics (cellophane and celluloid)
- nondrip paint
- whipped dessert toppings.

centre of gravity

The point in an object about which its **weight** is evenly balanced. In a uniform gravitational field, this is the same as the centre of **mass**.

centrifugal force

A useful concept in **physics**, based on an apparent (but not real) force. It may be regarded as a force that acts radially outward from a spinning or orbiting object, thus balancing the **centripetal force** (which is real).

- For an object of **mass** m moving with a **velocity** v in a circle of radius r, the centrifugal force F equals $\frac{mv^2}{r}$ (outward).

centripetal force

A force that acts radially inward on an object moving in a curved path. For example, with a weight whirled in a circle at the end of a length of string, the centripetal force is the tension in the string. The reaction to this force is the **centrifugal force**.

- For an object of **mass** m moving with a **velocity** v in a circle of radius r, the centripetal force F equals $\frac{mv^2}{r}$ (inward).

chain reaction

A **reaction** that is self-sustaining because the products of one step trigger the next step. A chain reaction is characterized, therefore, by the continual generation of reactive substances.

Stages in a chain reaction
- initiation – the initial generation of the reactants
- propagation – the reactants generate similar or different reactants
- termination – the reactants produce only stable, nonreactive substances.

Chain reactions may occur slowly (for example, the **oxidation** of edible oils) or accelerate as the number of reactants increases, ultimately resulting in an explosion.

Nuclear chain reaction
Neutrons released by the splitting of some atomic nuclei themselves go on to split others, releasing even more neutrons. Such a reaction can be controlled (as in a nuclear reactor) by using moderators to absorb excess neutrons. Uncontrolled, a chain reaction produces a nuclear explosion. The minimum amount of fissile material that can undergo a continuous chain reaction is referred to as the **critical mass**.

chemical equation
A chemical equation gives two basic pieces of information about a chemical **reaction**:
- the reactants (on the left-hand side) and products (right-hand side)
- how many units of each reactant and product are involved.

The equation must balance; that is, the total number of **atoms** of a particular **element** on the left-hand side must be the same as the number of atoms of that element on the right-hand side.

chemical kinetics
The branch of **chemistry** that investigates the rates of chemical reactions.

chemistry
The branch of science concerned with the study of the structure and composition of **matter**, the changes that matter may undergo and the phenomena that occur in the course of these changes.

The branches of chemistry
- *Organic chemistry* deals with the description, properties, reactions, and preparation of **carbon** compounds.

- *Inorganic chemistry* deals with all the other **elements** and the **compounds** they form.
- *Physical chemistry* is concerned with the quantitative explanation of chemical phenomena and **reactions**, and the measurement of data required for such explanations. This branch of chemistry studies in particular the movement of **molecules** and the effects of temperature and **pressure**, often with regard to **gases** and **liquids**.
- *Biochemistry* is the study of the chemistry of living things.

chemotherapy

Any medical treatment that involves the use of chemicals. It usually refers to treatment of **cancer** with cytotoxic (cell poisoning) and other **drugs**. The term was coined by the German bacteriologist Paul **Ehrlich** for the use of synthetic chemicals against infectious **diseases**.

CFC (chlorofluorocarbon)

A class of synthetic chemicals that are odourless, nontoxic, nonflammable, and chemically inert. The first CFC was synthesized in 1892, but no use was found for it until the 1920s. Subsequently their stability and apparently harmless properties made CFCs popular as propellants in aerosol cans, as refrigerants in refrigerators and air conditioners, as degreasing agents, and in the manufacture of foam packaging.

Ozone eaters
Under the influence of **ultraviolet radiation** from the Sun, CFCs in the atmosphere react with **ozone** (O_3) to form free chlorine (Cl) atoms and molecular **oxygen** (O_2), thereby destroying the ozone layer that protects Earth's surface from the Sun's harmful ultraviolet rays. The chlorine liberated can react with still more ozone, making the CFCs particularly dangerous to the **environment**. In June 1990 representatives of 93 nations, including the UK and the USA, agreed to phase out production of CFCs and various other ozone-depleting chemicals by the end of the 20th century.

chlorophyll

A green pigment responsible for the absorption of light **energy** during **photosynthesis**. Chlorophyll is found within **chloroplasts** which are present in large numbers in the **cells** of plant leaves, giving them their green colour. Cyanobacteria (blue-green algae) and other photosynthetic **bacteria** also have chlorophyll, though of a slightly different type.

chloroplast

A structure within a plant cell containing the green pigment **chlorophyll**. Chloroplasts occur in most cells of the green plant that are exposed to light, often in large numbers. Typically, they are flattened and disclike. They contain a small amount of **DNA** and divide by fission.

cholesterol

A white, crystalline substance found throughout the body, especially in **fats**, **blood**, nerve tissue, and bile; it is also provided in the diet by **foods** such as eggs, meat, and butter. Cholesterol is an integral part of all **cell** membranes and the starting point for the manufacture of steroid **hormones**, including the sex hormones. It is an essential component of lipoproteins, which transport fats and fatty acids in the blood.

- *Low-density lipoprotein cholesterol* (LDL-cholesterol), when present in excess, can become deposited on the surface of the arteries, causing atherosclerosis (hardening of the arteries).
- *High-density lipoprotein cholesterol* (HDL-cholesterol) acts as a scavenger, transporting fat and cholesterol from the tissues to the liver to be broken down.

Blood cholesterol levels can be altered by reducing the amount of alcohol and fat in the diet and by substituting some of the saturated fat for **polyunsaturated** fat, which brings about a reduction in LDL-cholesterol. HDL-cholesterol can be increased by exercise.

chromatography

A technique for separating or analysing a mixture of gases, liquids, or dissolved substances. There are several types of chromatography but basically all involve a mobile carrier substance (the mobile phase) transporting the sample mixture through another substance (the stationary phase). The mobile phase may be a **gas** or a **liquid**; the stationary phase may be a liquid or a **solid**, and may be in a column, on paper, or in a thin layer on a glass or plastic support. The components of the mixture are absorbed or impeded by the stationary phase to different extents and therefore become separated.

- *Paper chromatography*: the mixture separates because the components have differing solubilities in the solvent flowing through the paper and in the chemically bound water of the paper.
- *Thin-layer chromatography*: a wafer-thin layer of adsorbent medium on a glass plate replaces the filter paper. The mixture separates because of the

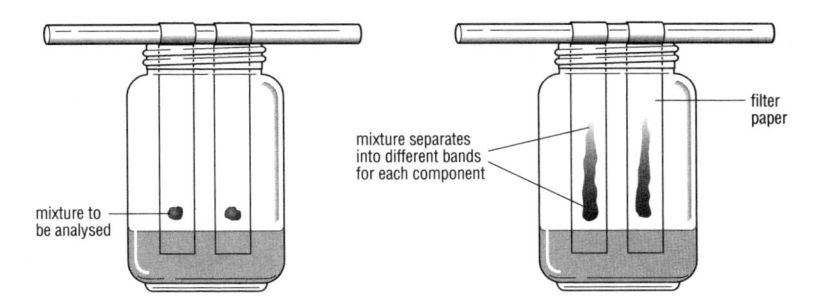

chromatography *Paper chromatography.*

differing solubilities of the components in the solvent flowing up the solid layer, and their differing tendencies to stick to the solid (adsorption).

- *Gas–liquid chromatography*: a gaseous mixture is passed into a long, coiled tube (enclosed in an oven) filled with an inert powder coated in a liquid. A carrier gas flows through the tube. The mixture separates as the components dissolve in the liquid to differing extents or stay as a gas.

chromosome

One of a number of threadlike structures found within in a cell **nucleus** during **cell division.** Chromosomes carry the **genes** that determine the characteristics of an organism. Chromosomes are only visible during cell division; at other times they exist in a less dense form called chromatin. Each chromosome consists of one very long strand of **DNA**, coiled and folded around a **protein** to produce a compact body. Most higher organisms have two copies of each chromosome, together known as a homologous pair. There are 23 pairs of chromosomes in a normal human **cell**. *(See illustration on p 50).*

> The first artificial human chromosome was constructed by US geneticists in 1997.

circadian rhythm

Any repeated activity that takes place over a 24-hour period, for example the sleep/activity cycle of animals. Circadian rhythms are generally controlled by **biological clocks.**

classification

The arrangement of organisms into a hierarchy of groups on the basis of their similarities. Evolutionary theory led to the development of phyloge-

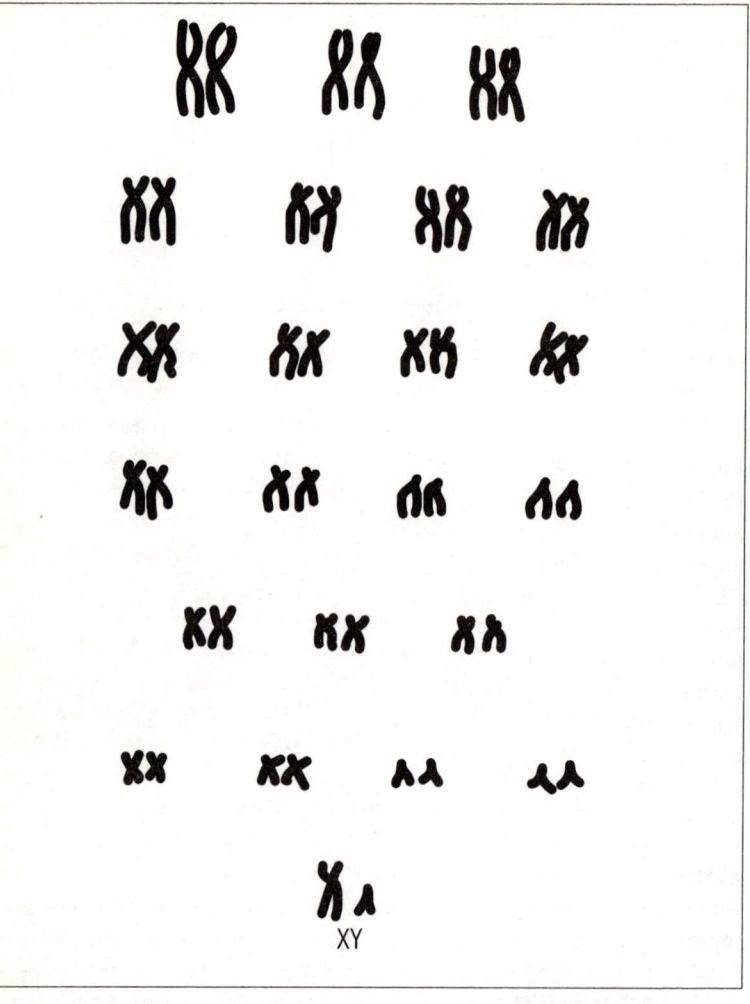

chromosome *The 23 pairs of chromosomes of a normal human male.*

netic classification, which aims to classify organisms in a way that mirrors their evolutionary and genetic relationship. Species are grouped according to shared characteristics believed to be derived from common ancestors (care being taken to exclude shared characteristics known to be due to convergent evolution).

classification hierarchy
Kingdom
    Subkingdom
Phylum (animals)/Division (plants)
    Subphylum
Class
    Subclass
    Infraclass
Order
    Superfamily
Family
Genus
Species
    Subspecies

Kingdom: Animalia (animals)
Phylum: Chordata (spinal cord)
Subphylum: Vertebrata (backbone)
Superclass: Tetrapoda (four limbs)
Class: Mammalia (suckling young)
Subclass: Theria (live births)
Infraclass: Eutheria (placenta)
Order: Primates (most highly developed)
Superfamily: Hominoidea (humanlike)
Family: Homonidae (two-legged)
Genus: *Homo* (human)
Species: *sapiens* (modern humans)

clone

An organism or **cell** that has arisen from another organism or cell by **asexual reproduction** and is therefore genetically identical to the original. In February 1997 scientists at the Roslin Institute in Scotland cloned an adult sheep from a single cell to produce Dolly, a lamb with the same **genes** as its mother. A cell was taken from the udder of the mother sheep, and its **DNA** combined with an unfertilized egg that had had its DNA removed. The fused cells were grown in the laboratory and then implanted into the uterus of a surrogate mother sheep. This was the first successful cloning of an adult **mammal**.

The term 'clone' has also been adopted by computer technology to describe a (nonexistent) device that mimics an actual one to enable certain software programs to run correctly.

colour

The sensation produced in the brain when **light** of different wavelengths falls on the eye. Sources of light have a characteristic **spectrum** or range of wavelengths. Visible white light consists of **electromagnetic radiation** of various wavelengths (if a beam is refracted through a prism, it spreads out into a spectrum, in which the various colours correspond to different wavelengths). When an object is illuminated by white light, some of the wavelengths are absorbed and some are reflected to the eye of an observer. The object appears coloured because of the wavelengths of light it reflects. For instance, plants are green because they absorb blue and red light and reflect green.

> **COLOUR VISION**
>
> Cone cells in the retina at the back of the eyeball are responsible for colour vision. There are three types, each sensitive to one colour only, either red, green, or blue. The brain combines the signals sent from the set of cones to produce a sensation of colour. Malfunctioning of the cone cells results in colour blindness.

community

A naturally occurring assemblage of **plants**, **animals**, and other organisms living within a certain area. Communities are usually named by reference to a dominant feature such as characteristic plant species (for example, a beechwood community), or a prominent physical feature (for example, a freshwater-pond community).

competition

The interaction between two or more organisms, or groups of organisms (for example, **species**), that use a common resource, such as food or water, that is in short supply in their **environment**. Competition invariably results in a reduction in the numbers of one or both competitors, and in **evolution** contributes both to the decline of certain species and to the evolution of **adaptations**.

compound

A chemical substance made up of two or more **elements** bonded together, so that they cannot be separated by physical means. Compounds are held together by ionic or covalent **bonds**. The proportions of the different elements in a compound are shown by the chemical formula of that compound. For example, a molecule of sodium sulphate, represented by the formula Na_2SO_4, contains two **atoms** of sodium, one of sulphur, and four of **oxygen**.

conduction, electrical
The flow of charged particles through a material giving rise to an **electric current**. Conduction in metals involves the flow of negatively charged free **electrons**. Conduction in **gases** and some **liquids** involves the flow of **ions** that carry positive charges in one direction and negative charges in the other.

conduction, thermal
The transmission of **heat** through a substance from a region of high **temperature** to a region of lower temperature. In **gases** and **liquids** heat is transferred by collisions between **atoms** in the material. In solids heat is transferred by the vibrations of atoms and **molecules**. In metals, which are excellent conductors, heat is carried rapidly by fast-moving **electrons**.

conductor
Any material that conducts **heat** or **electricity** (as opposed to an insulator, or nonconductor). A good conductor has a high electrical or heat conductivity, and is generally a substance rich in free **electrons** such as a **metal**. A poor conductor (such as the nonmetals, glass, and porcelain) has few free electrons.

corrosion
The eating away and eventual destruction of **metals** and **alloys** by chemical attack. The rusting of ordinary iron and steel is the most common form of corrosion.

cosmology
The branch of **astronomy** that deals with the structure and evolution of the **universe**. Cosmologists sometimes construct mathematical models of the universe and compare their properties with those of the observed universe to help them come up with theories about the nature and origin of the universe.
   See also: *Big Bang* for current views.

Crick, Francis Harry Compton (1916–)
English molecular biologist who, with James **Watson**, deciphered the molecular structure of **DNA**, and the means whereby characteristics are

CRICK'S LIFE

1916 Francis Crick is born in Nottingham, UK on 8 June.
1937 He gains a BSc in physics at University College, London.
1940 Crick works for the British Admiralty on the development of radar and magnetic mines for naval warfare.
1947 He joins the Cavendish Laboratory, Cambridge, laboratory to study biology, organic chemistry, and X-ray diffraction techniques.
1951 US biologist James Watson joins the laboratory and he and Crick form a close working relationship.
1953 Crick and Watson publish their work on the proposed structure of DNA.
1962 Crick and Watson share the Nobel Prize for Physiology or Medicine with Maurice Wilkins.
1977 Crick becomes a professor at the Salk Institute, San Diego, California.

transmitted from one generation to another. Crick later demonstrated that small segments of DNA, which he called codons, specified particular **amino acids** used for building **proteins**.

> ❝ We've discovered the secret of life! ❞
>
> **Francis Crick**, 28 February 1953

critical mass

The minimum mass of fissile radioactive material (see **fission**) that can undergo a continuous **chain reaction**. Below this mass, too many **neutrons** escape from the surface for a chain reaction to carry on; above the critical mass, the **reaction** may accelerate into a nuclear explosion.

cryogenics

The science of very low **temperatures**, including the means of producing such temperatures and the exploitation of special properties associated with them, such as the disappearance of electrical resistance (**superconductivity**). At temperatures near **absolute zero**, materials can display unusual properties. Some metals, such as mercury and lead, exhibit superconductivity. Liquid helium loses its **viscosity** and becomes a 'superfluid' when cooled to below 2 K; in this state it flows up the sides of its container.

> **PRACTICAL CRYOGENICS**
>
> Cryotherapy is a process used in eye surgery, in which a freezing probe is briefly applied to the outside of the eye to repair a break in the retina.

crystal

A substance in which the **atoms** or **molecules** are arranged in an orderly three-dimensional pattern, resulting in an external surface of clearly defined smooth faces with characteristic angles between them. A **mineral** can often be identified by the shape of the crystals it forms. Table **salt** and quartz are examples of crystals.

> A single crystal can vary in size from a submicroscopic particle to a mass some 30 m/100 ft in length.

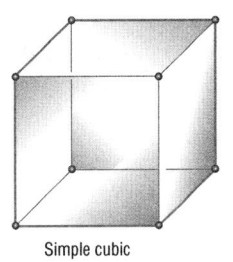

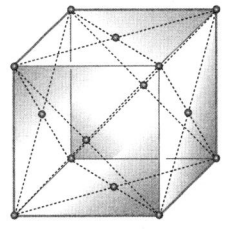

Simple cubic

Face-centered cubic

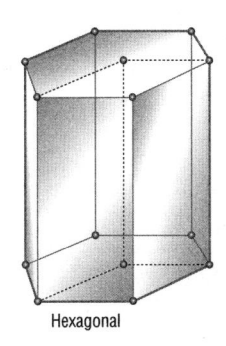

Hexagonal

crystal *Crystals are classified into systems by how symmetrical they are. Cubic crystals are symmetrical from almost any angle.*

crystallography

The scientific study of crystals. In 1912 it was found that the shape and size of the repeating atomic patterns (unit cells) in a crystal could be determined by passing **X-rays** through a sample. This method, known as X-ray diffraction, opened up an entirely new way of 'seeing' atoms. It has been found that many substances have a unit cell that exhibits all the symmetry of the whole crystal; in table salt (sodium chloride, NaCl), for instance, the unit cell is an exact cube.

Curie, Marie, born Manya Skłodowska, (1867–1934) **and Pierre** (1859–1906)
Polish and French scientists who, in 1898, discovered two new radioactive **elements**: polonium and radium. Both scientists were jointly awarded the Nobel Prize for Physics in 1903, which they shared with Henri Becquerel. Marie Curie was also awarded the Nobel Prize for Chemistry in 1911. The Curies took no precautions against **radioactivity**. Marie Curie's notebooks are too contaminated to handle, even today.

Curie *Marie and Pierre Curie with their eldest daughter Irene in 1904.*

LIVES OF THE CURIES

1859	Pierre Curie is born in Paris, France on 15 May.
1867	Manya Sklodowska is born in Warsaw, Poland on 7 November.
1878	Pierre becomes an assistant at the Sorbonne.
1891	Marie enters the Sorbonne and studies physics and mathematics, graduating top of her class.
1895	Marie marries French chemist Pierre Curie. Pierre discovers the Curie point, the critical temperature at which a paramagnetic substance become ferromagnetic.
1896	The Curies begin working together on radioactivity.
1898	They announce the discovery of polonium, followed by that of radium.
1903	The Curies are awarded the Nobel Prize for Physics for their discovery of polonium and radium.

1906	Pierre Curie is run down and killed by a horse-drawn carriage. Marie becomes the first woman to teach at the Sorbonne when she takes over his teaching.
1910	Marie isolates pure radium metal with André Debierne (1874–1949).
1911	Marie Curie is awarded the Nobel Prize for Chemistry for her work on the chemistry and medical applications of radium.
1914	At the outbreak of World War I Marie Curie helps to equip ambulances with X-ray equipment, which she drives to the front lines.
1934	Marie Curie dies of leukaemia.

cybernetics

The science of systems and how they organize, regulate, and reproduce themselves, and also how they evolve and learn. In the laboratory, inanimate objects are created that behave like living systems. Applications range from the creation of electronic artificial limbs to the running of the fully automated factory where decision-making machines operate up to managerial level.

Cybernetics was founded and named in 1947 by US mathematician Norbert Wiener. Originally, it was the study of control systems using feedback to produce automatic processes.

D

Dalton, John (1766–1844)
English chemist who proposed the theory of **atoms**, which he considered to be the smallest parts of **matter**. From experiments with **gases**, Dalton noted that the proportions of two components combining to form another gas were always constant. He suggested that if substances combine in simple numerical ratios, then the **weight** proportions represent the relative atomic masses of those substances. He also proposed the law of partial **pressures** (Dalton's law), stating that for a mixture of gases the total pressure is the sum of the pressures that would be developed by each individual gas if it were the only one present.

dark matter
According to current theories of **cosmology**, over 90% of the **mass** of the **universe** has so far remained undetected. This dark matter, if shown to exist, would explain many currently unexplained gravitational effects in the movement of **galaxies**. Theories of the composition of dark matter include unknown atomic particles (cold dark matter) or fast-moving neutrinos (hot dark matter) or a combination of both. Approximately half of the dark matter surrounding the Milky Way could be accounted for by unseen astronomical objects such as burned out stars, stars that failed to light up, and planets.

BACKWARD TIME UNIVERSE

In 1999 scientists put forward an extraordinary proposal to explain dark matter. Anti-time particles travelling backwards in time from the collapsing universe of the far future collide with normal time particles, leaving them in an undetectable state of non-time but retaining their gravitational effect!

Darwin, Charles Robert (1809–1882)

English naturalist who developed the modern theory of **evolution** and proposed, with Alfred Russel **Wallace**, the principle of **natural selection**.

Darwin's work marked a turning point in many of the sciences, including physical **anthropology** and palaeontology. Between 1831 and 1836, Darwin travelled as a naturalist aboard HMS *Beagle*. In South America, he saw **fossil** remains of giant sloths and other animals now **extinct**, and on the Galápagos Islands he found 14 similar species of finch, none of which existed on the mainland (see **adaptation**). It was apparent that one ancestor finch must have evolved into the others, but how it did so eluded him. In 1859 Darwin published *On the Origin of Species by Means of Natural Selection or the Preservation of Favoured Races in the Struggle for Life*. This

DARWIN'S LIFE

1809	Darwin is born in Shrewsbury, UK on 12 February.
1825	He studies medicine at Edinburgh University.
1827	Darwin's father sends him to study Theology at Christ's College Cambridge.
1831–1836	Darwin accompanies HMS Beagle as a naturalist on its survey voyage of the coasts of Patagonia, Tierra del Fuego, Chile, Peru, and some Pacific islands.
1838	He is appointed secretary to the Geological Society.
1839	He publishes his findings from the voyage in the book: *Journal of Researches into the Geology and Natural History of the Various Countries Visited by HMS Beagle (1832–1836)*.
1856	Darwin begins writing fully about evolution and natural selection.
1858	He receives a paper from fellow naturalist, Alfred Wallace, explaining exactly the same theory of evolution and natural selection.
1859	He publishes an abstract of his work entitled *The Origin of the Species by Means of Natural Selection or the Preservation of Favoured Races in the Struggle for Life*.
1871	Darwin publishes *The Descent of Man and Selection in Relation to Sex*.
1882	He dies on 19 April at Down, in Kent.

book explained the evolutionary process through the principles of natural selection and aroused bitter controversy because it disagreed with the literal interpretation of the Book of Genesis in the Bible. It was not until his publication of *The Descent of Man and Selection in Relation to Sex* (1871), that Darwin argued that people had evolved just like other organisms.

The key work on **heredity** by the Austrian scientist Gregor **Mendel** was carried out during Darwin's own lifetime and published in 1865, but was neglected until 1900.

> ❛ What can be more curious than that the hand of a man, formed for grasping, that of a mole for digging, [...] and the wing of a bat, should all be constructed on the same pattern, and should include the same bones, in the same relative positions? ❜
>
> **Charles Darwin**, *On the Origin of Species*, 1859

decomposition

The process whereby a chemical **compound** is reduced to its component substances.

- *Thermal decomposition* occurs as a result of heating.
- *Electrolytic decomposition* may result when an electrical current is passed through a compound in the molten state or in aqueous solution.
- *Catalysed decomposition* occurs when decomposition is aided by the presence of a catalyst.

In **biology**, decomposition is the breaking down of dead organisms either chemically or by the action of decomposers, such as **bacteria** and **fungi**.

dendrochronology

Dating past events by analysing the annual rings of trees to determine the age of timber. Since annual rings are formed by variations in the water-conducting cells produced by the plant during different seasons of the year, they also provide a means of establishing past climatic conditions in a given area. Samples of wood are obtained by driving a narrow metal tube into a tree to remove a core extending from the bark to the centre. Samples taken from timbers at an archaeological site can be compared with a master core on file for that region or by taking cores from old living trees; the year when they were felled can be determined by locating the point where the rings of the two samples correspond and counting back from the present.

> **ANCIENT HISTORY**
>
> Sequences of tree rings extending back over 8,000 years have been obtained for the southwest USA and northern Mexico by using cores from the bristlecone pine *Pinus aristata* of the White Mountains, California, which can live for over 4,000 years in that region.

density

A measure of the compactness of a substance; it is equal to its **mass** per unit **volume** and is measured in kg per cubic metre/lb per cubic foot.

- The average density D of a mass m occupying a volume V is given by the formula:

$$D = \frac{m}{V}$$

- Relative density is the ratio of the density of a substance to that of water at 4°C/32.2°F.

Densities of some common substances

Substance	Density in kg m^{-3}	Substance	Density in kg m^{-3}
Solids		lead	11,300
balsa wood	200	uranium	19,000
oak	700		
butter	900	*Liquids*	
ice	920	water	1,000
ebony	120	petrol, paraffin	800
sand (dry)	1,600	olive oil	900
concrete	2,400	milk	1,030
aluminium	2,700	sea water	1,030
steel	7,800	glycerine	1,260
copper	8,900	Dead Sea brine	1,800

Densities of some common substances (*continued*)

Substance	Density in kg m^{-3}	Substance	Density in kg m^{-3}
Gases		helium	0.18
(at standard temperature		methane	0.72
and pressure of 0°C		nitrogen	1.25
and 1 atm)		oxygen	1.43
air	1.30	carbon dioxide	1.98
hydrogen	0.09	propane	2.02

diffraction

The spreading out of **waves** when they pass through a small gap or around a small object, resulting in some change in the direction of the waves. The size of the object or gap must be comparable to or smaller than the wavelength of the waves for diffraction to occur.

- All forms of waves – **electromagnetic**, **sound**, and water waves – can be diffracted. Long-wave radio waves can bend round hills, for example. The slight spreading of a light beam through a narrow slit causes the different wavelengths of light to interfere with each other to produce a pattern of light and dark bands.

HEARING ROUND CORNERS

When sound waves travel through doorways or between buildings they are diffracted significantly, so that the sound can be heard round corners.

---

diffusion

The random movement of **molecules** or particles in a fluid (**gas** or **liquid**) from a region of high concentration to a region of lower concentration, until a uniform concentration is achieved throughout. The difference in concentration between two such regions is called the concentration gradient. No mechanical mixing or stirring is involved. For instance, if a drop of ink is added to water, its molecules will diffuse until their colour becomes evenly distributed throughout.

In biological systems, diffusion plays an essential role in the transport, over short distances, of molecules such as nutrients, respiratory gases, and neurotransmitters. It provides the means by which small molecules pass into and out of individual **cells** and **micro-organisms**, such as amoebae, that possess no circulatory system. Organs such as the lungs, whose function depends on diffusion, have a large surface area.

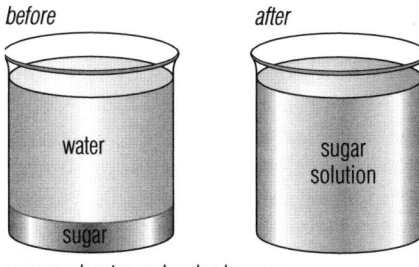

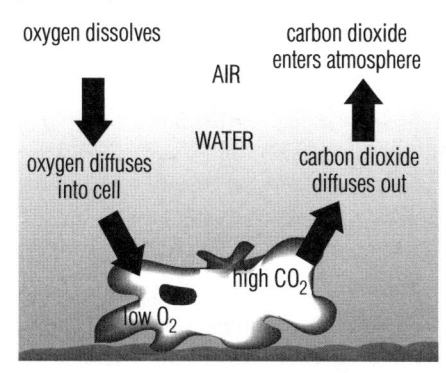

before / *after*

sugar and water molecules become evenly mixed

gas exchange in amoeba

diffusion *Diffusion is the movement of molecules from a region of high concentration into a region of lower concentration.*

digestive system

All the organs and tissues involved in the digestion of food. In **animals**, these consist of the mouth, stomach, intestines, and their associated glands. The process of digestion breaks down the food by physical and chemical means into the different elements that are needed by the organism for **energy** and tissue building and repair.

dimension

In science, a dimension is any directly measurable physical quantity such as **mass** (M), length (L), and **time** (T), and the units that are derived from such quantities by multiplication or division. For example, acceleration (the rate of change of **velocity**) has dimensions (LT), and is expressed in such units as kms^{-2}. A quantity that is a ratio, such as relative density or humidity, is dimensionless.

In geometry, the dimensions of a figure are the number of measures needed to specify its size. A point is considered to have zero dimension, a line to have one dimension, a plane figure to have two, and a solid body to have three.

dinosaur

Any of a group of extinct **reptiles** living between 205 million and 65 million years ago. Their closest living relations are crocodiles and **birds**. Many species of dinosaur evolved during the millions of years they were the dominant large land animals. Many were large (up to 27 m/90 ft), but some were as small as chickens. They disappeared 65 million years ago for reasons not fully understood, although many theories exist, perhaps the most widely accepted being that Earth was struck by a comet.

Dinosaurs are divided into the orders Saurischia ('lizard-hip') and Ornithischia ('bird-hip'). Members of the former group possessed a reptile-like pelvis and were mostly bipedal and **carnivorous**, although some were giant amphibious quadrupedal **herbivores**. Members of the latter group had a bird-like pelvis, were mainly four-legged, and entirely herbivorous.

disease

A disorder in the normal functioning of an organism. Diseases can occur in all life forms, and can affect **cells**, tissues, organs, or systems. Diseases are usually characterized by specific symptoms and signs, and can be mild and short-lasting – such as the common cold – or severe enough to wipe out a whole **population** – such as Dutch elm disease.

- *Infectious diseases* are caused by **micro-organisms**, such as **bacteria** and **viruses**, invading the body. They can be spread across a species, or transmitted between one or more **species**.

- *Noninfectious diseases* may be inherited (congenital diseases); caused by the ingestion or absorption of harmful substances, such as **toxins**; result from poor nutrition or hygiene; or may arise from injury or ageing.

DNA (deoxyribonucleic acid)

DNA is the recipe book for making an organism. It is a complex giant molecule that contains, in chemically coded form, the information needed for a **cell** to make **proteins**. It forms the basis of genetic inheritance in all organisms, except for a few **viruses** that use **RNA** (ribonucleic acid) to carry their genetic information.

DNA is a double-stranded nucleic acid made up of two chains of subunits called nucleotides. Each nucleotide contains either a purine (adenine or guanine) or pyrimidine (cytosine or thymine) base. These bases link up with each other in a specific way: adenine links with thymine; cytosine with guanine to form base pairs that connect the two strands of the DNA

DNA · 65

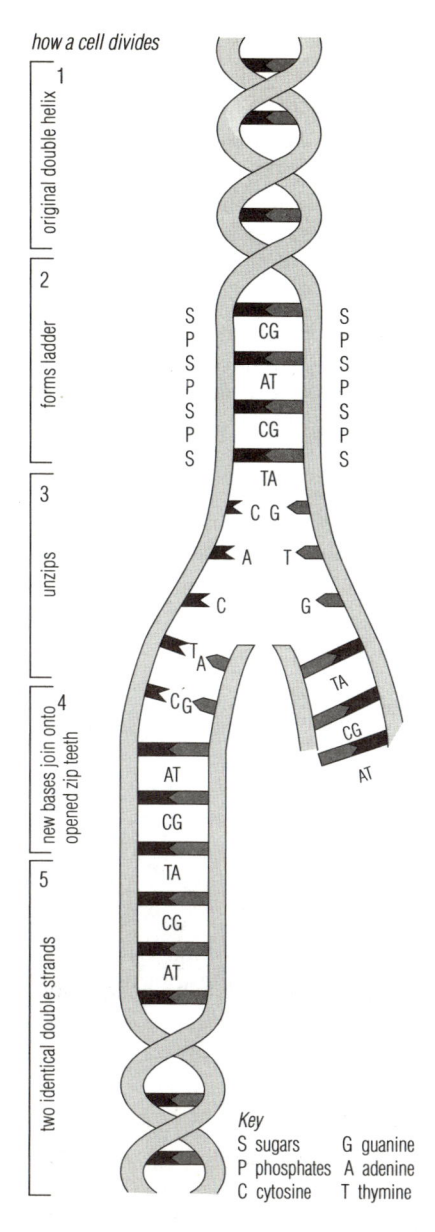

DNA *How the DNA molecule divides. The DNA molecule consists of two strands wrapped around each other in a spiral or helix. The main strands consist of alternate sugar (S) and phosphate (P) groups, and attached to each sugar is a nitrogenous base – adenine (A), cytosine (C), guanine (G), or thymine (T). The sequence of bases carries the genetic code which specifies the characteristics of offspring. The strands are held together by weak bonds between the bases, cytosine to guanine, and adenine to thymine. The weak bonds allow the strands to split apart, allowing new bases to attach, forming another double strand.*

molecule like the rungs of a twisted ladder.

Hereditary information is stored as a specific sequence of bases. A set of three bases – known as a codon – represents a particular **amino acid**, the subunit of a protein molecule. The codons are read and transcribed by RNA and translated into protein production. Because proteins are the chief structural molecules of living matter and, as **enzymes**, regulate all aspects of **metabolism**, it may be seen that the **genetic code** is effectively responsible for building and controlling the whole organism.

If all the DNA in one human body was laid out in a line it would stretch over 160 billion km/100 billion mi.

Doppler effect
The change in pitch heard in a **sound** as the source moves away or comes nearer. The Doppler effect is responsible for the perceived change in pitch of a siren as it approaches and then recedes. Doppler shift also affects light **waves** and accounts for the shift of light towards the red end of the **spectrum** from distant galaxies that are moving away from us. It is named after the Austrian physicist Christian Doppler.

dormancy
A phase of reduced physiological activity exhibited by certain buds, seeds, and spores. Dormancy can help a plant to survive unfavourable conditions, as in annual plants that pass the cold winter season as dormant seeds, and plants that form dormant buds.

drug
Any of a range of substances, natural or synthetic, administered to humans and animals as therapeutic agents: to diagnose, prevent, or treat **disease**, or to assist recovery from injury. Traditionally many drugs were obtained from plants or animals; some **minerals** also had medicinal value. Today, increasing numbers of drugs are synthesized in the laboratory.

WHAT'S IN A NAME?

Drugs generally have three names. The first is the chemical name, which is often too complicated to remember. Then there is the generic or non-proprietary name given when the drug is approved for medical use. Such a drug may have BP (British Pharmacopoeia) or USP (United States Pharmacopoeia) after its name. Thirdly there is the proprietary, or trade, name given to the drug by the manufacturing company that initially takes out a patent on their synthesis. One compound may have a large number of proprietary names.

---

dye
A substance that is applied to fabrics to give them colour. Direct dyes combine with the material of the fabric, yielding a coloured compound; indirect dyes require the presence of another substance (a mordant), with which the fabric must first be treated; vat dyes are colourless soluble substances that on exposure to air yield an insoluble coloured compound.

E

ear

The organ of hearing in animals which responds to the vibrations that constitute sound. The vibrations are translated into nerve signals and passed to the brain. A mammal's ear consists of three parts: outer ear, middle ear, and inner ear. The outer ear is a funnel that collects sound, directing it down a tube to the ear drum (tympanic membrane). Sounds vibrate this membrane, the mechanical movement of which is transferred to a smaller membrane leading to the inner ear by three small bones, the auditory ossicles. Vibrations of the inner ear membrane then move fluid contained in the snail-shaped cochlea, which in turn vibrates hair cells that stimulate the auditory nerve connected to the brain.

> When a loud noise occurs, muscles behind the eardrum contract automatically, suppressing the noise to enhance perception of sound and prevent injury.

echo

The repetition of a **sound**, or of a radar or sonar signal, by the **reflection** of waves from a surface. By accurately measuring the time taken for an echo to return to the transmitter, and by knowing the speed of a radar signal (the **speed of light**) or a sonar signal (the speed of sound in water), it is possible to calculate the range of the object causing the echo (**echolocation**). A similar technique is used in echo sounders to estimate the depth of water under a ship's keel or the depth of a shoal of fish.

echolocation

A method used by certain **animals**, notably bats, whales, and dolphins, to detect the positions of objects by using sound. The animal emits a stream of high-pitched sounds, generally at frequencies beyond the range of human hearing, and listens for the returning echoes reflected off objects to determine their exact location.

The location of an object can be established by the time difference between the arrival of the **echo** at each ear. Echolocation is of particular value under conditions when normal vision is poor (at night in the case of bats, in murky water for dolphins). A few species of **bird** can also echolocate, including cave-dwelling birds such as some species of swiftlets and the South American oil bird.

ecology

The study of the relationships between organisms and the **environments** in which they live, including all living and nonliving components. The chief environmental factors governing the distribution of **plants** and **animals** are **temperature**, **humidity**, soil, **light** intensity, daylength, **food** supply, and interaction with other organisms.

Ecology may be concerned with individual organisms (for example, behavioural ecology, feeding strategies), with **populations** (for example, population dynamics), or with entire communities (for example, competition between species for access to resources in an **ecosystem**, or predator–prey relationships). Applied ecology is concerned with the management and conservation of **habitats** and the consequences and control of pollution.

> **A PLACE TO LIVE**
>
> The term ecology was coined by the biologist Ernst Haeckel in 1866 from the Greek for house.

ecosystem

An integrated unit consisting of a **community** of living organisms and their shared physical environment. Individual organisms interact with each other and with their **environment**, or **habitat**, in a series of relationships that depends on the flow of **energy** and nutrients through the system. These relationships are usually complex and finely balanced, and in theory natural ecosystems are self-sustaining. Major changes to an

> **WORLD'S OLDEST ECOSYSTEM**
>
> In 1999, palaeontologists discovered fossilised stromatolites that have been dated at 3.46 billion years old near the town of Marble Bar in Western Australia. That would make this ancient colony of **bacteria** the world's oldest ecosystem.

ecosystem, such as climate change, overpopulation, or the removal of a **species**, may threaten the system's sustainability and result in its eventual destruction. For instance, the removal of a major **carnivore** predator can result in the destruction of an ecosystem through overgrazing by **herbivores**.

Ecosystems can contain smaller systems and be contained within larger ones. The global ecosystem (the ecosphere) consists of all the Earth's physical features – its land, oceans, and atmosphere (the geosphere) – together with all the biological organisms living on Earth (the biosphere). On a smaller scale, a freshwater-pond ecosystem includes the plants and animals living in the pond, the pond water and all the substances dissolved or suspended in it, together with the rocks, mud, and decaying matter at the bottom of the pond.

Ehrlich, Paul (1854–1915)
German bacteriologist and immunologist who pioneered **chemotherapy**, the treatment of **disease** using chemicals. He shared the 1908 Nobel Prize for Physiology or Medicine with Ilya Mechnikov, awarded for his work on **immunity**. Ehrlich was one of the earliest workers on immunology, and through his studies on blood samples the discipline of haematology was recognized.

> ❛ Success in research needs four Gs: *Glück, Geduld, Geschick und Geld.*
> Luck, patience, skill, and money. ❜
>
> **Paul Ehrlich**, quoted in M Perutz, *Nature*, 1988

Einstein, Albert (1879–1955)
German-born US physicist whose theories of **relativity** brought about a revolution in our understanding of **matter**, space, and **time**. Einstein also left his mark in other areas of physics and was awarded the Nobel prize for his work on the photoelectric effect.

Brownian motion
Einstein's first major achievement concerned Brownian movement, the random movement of fine particles first observed in 1827 by Scottish botanist Robert Brown (1773–1858). Einstein explained this phenomenon as being

the effect of large numbers of **molecules** bombarding the particles. He was able to make predictions of the movement and sizes of the particles, which were later verified experimentally. Einstein's explanation of Brownian motion was one of the most important pieces of evidence for the hypothesis that matter is composed of **atoms**.

❛ The most incomprehensible thing about the world is that it is at all comprehensible. ❜

Albert Einstein

The photoelectric effect
Light striking the surface of certain **metals** causes **electrons** to be emitted – the photoelectric effect. It had been found experimentally that electrons are not emitted by light of less than a certain **frequency** and that the **energy** of the electrons emitted increases with an increase in the frequency of the light. Einstein suggested that light energy behaved as if it were particles, which he called called 'light quanta' (later called **photons**). A photon can only release an electron if it has sufficient energy to do so. The higher the frequency the more energy it has.

MASS AND ENERGY

In 1907 Einstein showed that **mass** is related to energy by the famous equation $E=mc^2$, energy (E) equals mass (m) times the **speed of light** (c) squared. As the speed of light is a very large number this indicates the enormous amount of energy that is stored as mass. Some of it is released in **radioactivity** and nuclear reactions, for example in the Sun.

EINSTEIN'S LIFE

1879	Einstein is born on March 14 in Ulm, Germany.
1901	He becomes a Swiss citizen.
1909	He becomes a junior professor at the University of Zurich.

1911	Einstein is offered the post of Professor in Prague and then in Zurich, a year later.
1914	He is appointed Director of the Institute of Physics at the Kaiser Wilhelm Institute in Berlin.
1915	He publishes his general theory of relativity in which he predicted that light rays are bent by gravity, confirmation of this prediction, during the solar eclipse of 1919, makes him world famous.
1921	He is awarded the Nobel Prize for Physics for his work on photo-electricity.
1933	Einstein emigrates to the USA and accepts a position at the Princeton Institute for Advanced Study.
1940	He becomes a US citizen.
1952	The state of Israel pays him the highest honour it can by offering him the country's presidency, which he does not accept.
1955	Einstein dies on 18 April in Princeton, NJ, USA.

elasticity

The ability of a material to return to its original shape once a **stress** has been removed. An elastic material obeys **Hooke's** law, which states that its deformation is proportional to the applied stress up to a certain point, called the elastic limit, beyond which additional stress will deform it permanently. Elastic materials include **metals** and rubber; however, all materials have some degree of elasticity.

electric charge

A property of some elementary particles that causes them to exert forces on each other that give rise to the phenomena associated with **electricity**. Charge occurs in two forms, either negative or positive. Two bodies, both with positive or both with negative charges, repel each other, whereas bodies with opposite or 'unlike' charges attract each other.

In atoms, **electrons** possess a negative charge, and **protons** an equal positive charge. In a neutral **atom** these charges are balanced. A body can be charged by **friction**, induction, or chemical change. The charge can be either an accumulation of electrons (negative charge) or loss of electrons (positive charge). Atoms can sometimes gain electrons to become negative **ions** or lose them to become positive ions.

electric current

The flow of electrically charged particles, usually free **electrons**, through a conductor. Current carries electrical **energy** from a power supply, such as a battery of **electrical cells**, to the components of the circuit, where it is converted into other forms of energy, such as **heat**, **light**, or motion. For charge to flow in a circuit there must be a potential difference (pd) applied across the circuit. This may be supplied in the form of a **battery** that has a positive terminal and a negative terminal. The electrons are repelled from the negative terminal side of the battery and attracted to the positive terminal of the battery through the circuit. A steady flow of electrons around the circuit is produced.

- *Alternating current (AC):* is a current that flows in alternately reversed directions through or around a circuit. Electric energy is usually generated as alternating current in a power station, and alternating currents may be used for both power and lighting. The voltage of alternating current can be raised or lowered economically by a transformer: high voltage for generation and transmission, and low voltage for safe utilization. Railways, factories, and domestic appliances, for example, use alternating current.

- *Direct current (DC):* flows in one direction, and does not reverse its flow as alternating current does. The electricity produced by a battery is direct current.

electricity

All phenomena caused by **electric charge**, whether static or in motion. The first artificial electrical phenomenon to be observed was the ability of some naturally occurring materials such as amber to attract small objects such as dust and pieces of paper, when rubbed with a piece of cloth. When the amber is rubbed **electrons** are transferred from the cloth to the amber so

---

STEP UP, STEP DOWN

Electricity is generated at power stations at a voltage of about 25,000 volts. For minimal power loss, transmission takes place at very high voltage (400,000 volts or more). The generated voltage is 'stepped up' by a transformer and fed into the grid system, an interconnected network of power stations and distribution centres covering a large area. The line voltage is reduced by a step-down transformer and distributed to consumers.

that the amber becomes negatively charged. This accumulation of charge is called **static electricity**. Positive charges on the paper are attracted to the negative charges on the amber.

In a circuit the **battery** provides **energy** to make charge flow through the circuit as an **electric current**. The amount of energy supplied to each unit of charge is called the electromotive force (emf). The unit of emf is the volt (V).

See also: *electrolysis, electromagnetism, resistance.*

electrode

A conductor that emits or collects **electrons** in an **electric cell**; for example, the **anode** (positive electrode) or **cathode** (negative electrode) in a **battery**, or the carbons in an arc lamp.

electrolysis

Chemical changes brought about by passing an **electric current** through a **solution** or molten salt (the electrolyte), resulting in the movement of **ions** to the **electrodes**. Positive ions (cations) go to the negative electrode (**cathode**) and negative ions (anions) to the positive electrode (**anode**). During electrolysis, the ions react with the electrode, either receiving or giving up **electrons**. The resultant **atoms** may be liberated as a **gas**, or deposited as a **solid** on the electrode, in amounts that are proportional to the amount of current passed.

Applied electrolysis

- Electroplating, in which a solution of a salt, such as silver nitrate ($AgNO_3$), is used. The object to be plated acts as the negative electrode, thus attracting silver ions (Ag^+).

- Industrial processes, such as coating metals for vehicles and ships, and refining bauxite into aluminium.

  See also: *Faraday, Michael.*

electromagnetism

A coil of wire wound around a soft iron core that acts as a magnet when an **electric current** flows through the wire. Electromagnets have many uses: in switches, electric bells, solenoids and metal-lifting cranes. The strength of the electromagnet can be increased by increasing the current through the wire or by increasing the number of turns in the wire coil. At the north pole of the electromagnet current flows anticlockwise; at the south pole flow is clockwise.

Electromagnetism and the electric motor

In 1821, only one year after Danish physicist Hans Oersted (1777–1851) had discovered that a current of electricity flowing through a wire produces a magnetic field, Michael **Faraday** carried out experiments with magnets and wires arranged in such a way that either the magnets or the wires were free to move. When a current was passed through a freely moving wire it would spin round a fixed magnet, or a freely moving magnet would turn round a fixed wire. This experiment demonstrated the basic principles governing the electric motor.

electromagnetism *When an electric current passes through a wire wound round a magnet it induces movement in the magnet. The direction of the movement depends upon the direction of the electric current.*

Faraday went on to show that an electric current was produced when a magnet was moved near a coil or a coil moved near a magnet. This was the basis of the generator.

Electromagnetic induction and the transformer

In a series of experiments beginning in 1831 Faraday wound two coils, each insulated from the other, around an iron ring and connected one to a **battery** and the other to a galvanometer (an instrument for measuring electric currents). Faraday noticed that the galvanometer detected a current in the second wire whenever the current was switched on or off in the first. This current was caused, or induced, by variations in the magnetic field produced by the current in the first wire. With this device, Faraday had discovered the transformer.

electromagnetic spectrum

The complete range, over all wavelengths and **frequencies**, of **electromagnetic** waves. These range from long wavelength **radio** and television waves,

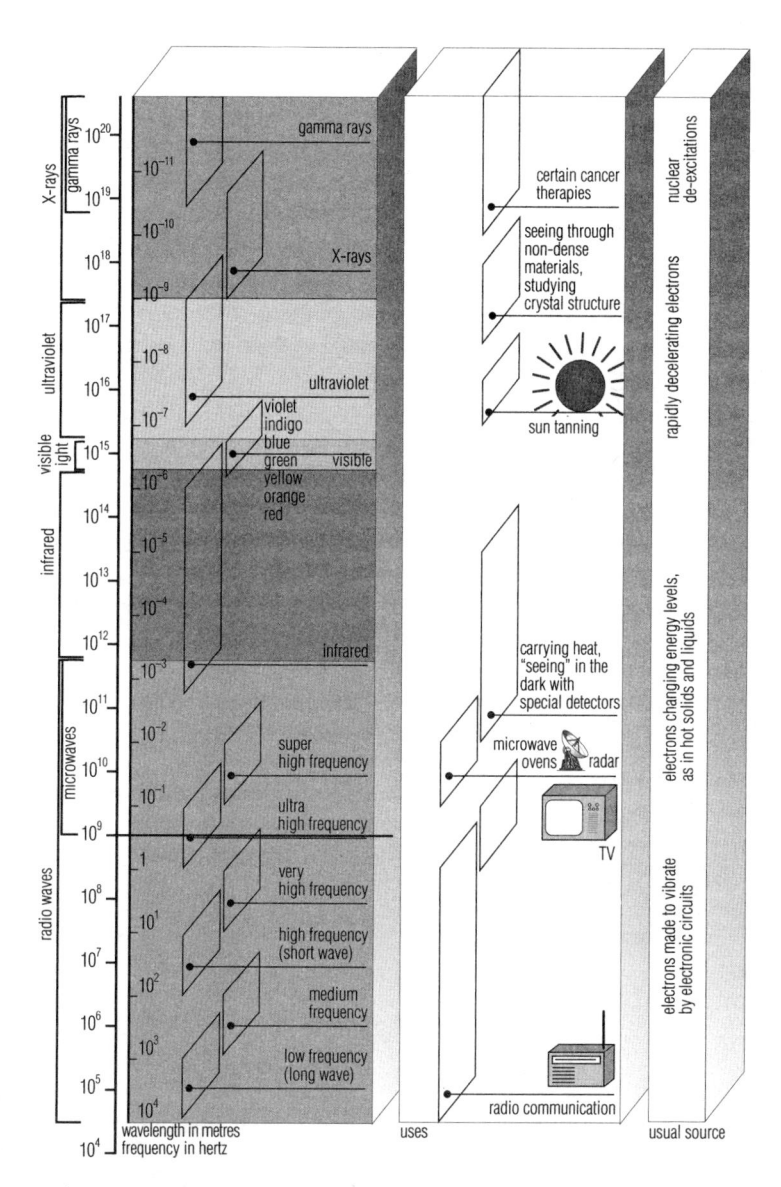

electromagnetic spectrum *Radio waves have the lowest frequency. Infrared radiation, visible light, ultraviolet radiation, X-rays, and gamma rays have progressively higher frequencies.*

infrared radiation, visible light, ultraviolet light, **X-rays**, and gamma radiation, the shortest wavelength and highest energy electromagnetic wave.

- Electromagnetic waves are oscillating electric and magnetic fields travelling together through space at a speed of nearly 300,000 km/186,000 mi per second (the **speed of light**).

electron

A stable, negatively charged elementary particle that is a constituent of all atoms, and a member of the class of particles known as leptons (see **particle physics**). Electrons surround the nucleus of an **atom** in layers, or shells; in a neutral atom the number of electrons is equal to the number of **protons** in the **nucleus**.

element

A substance that cannot be split chemically into simpler substances. Elements are classified in the **periodic table** of the elements. All the atoms of a particular element have the same number of **protons** and **electrons**, although the number of **neutrons** can vary (see **isotope**). Of the known elements, 92 are known to occur in nature. Others have been produced artificially in particle accelerators.

Forging the elements
According to current theories, hydrogen and helium were produced as subatomic particles condensed some time after the **Big Bang**. Elements up to atomic number 26 (iron) are made by nuclear fusion within the stars. The more massive elements, such as lead and uranium, are produced when an old star explodes; as its centre collapses, the gravitational energy squashes nuclei together to make new elements.

A BRIEF LIFE

After firing 5 billion billion zinc ions at a speed of 30,000 kps/18,640 mps at lead, German scientists created a single atom of element 112 that survived for a third of a millisecond in 1996.

---

embryo

An **animal** or a **plant** at the earliest stages of development after an egg **cell** begins to divide following **fertilization**, or activation by **parthenogenesis**. In

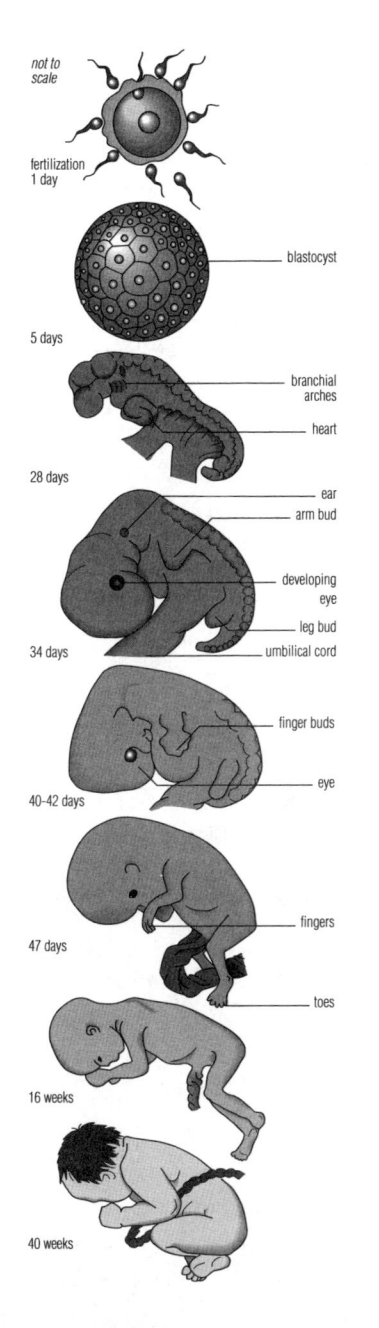

In humans, the term embryo describes the fertilized egg during its first seven weeks of existence; from the eighth week onwards it is referred to as a fetus.

---

animals the embryo exists either within an egg (where it is nourished by food contained in the yolk), or in **mammals**, in the uterus of the mother, where (except in marsupials) the embryo is fed through the placenta.

The plant embryo is found within the **seed** in higher plants. It sometimes consists of only a few cells, but usually includes a root, a shoot (or primary bud), and one or two cotyledons (seed leaves), which nourish the growing seedling.

embryology

The study of the changes undergone by an organism from conception to

embryology *The development of a human embryo. Division of the fertilized egg, or ovum, begins within hours of conception. Within a week a ball of cells –a blastocyst – has developed. After the third week, the embryo has changed from a mass of cells into a recognizable shape. At four weeks, the embryo is 3 mm/0.1 in long, with a large bulge for the heart and small pits for the ears. At six weeks, the embryo is 1.5 cm/0.6 in with a pulsating heart and ear flaps. At the eighth week, the embryo is 2.5 cm/1 in long and recognizably human, with eyelids, small fingers, and toes. From the end of the second month, the embryo is almost fully formed and further development is mainly by growth. After this stage, the embryo is termed a fetus.*

hatching or birth. It is mainly concerned with the changes in **cell** organization in the embryo and the way in which these lead to the structures and organs of the adult (the process of differentiation). Applications of embryology include embryo transplants, both commercial (for example, in building up a prize dairy herd quickly at low cost) and in obstetric medicine (as a method for helping couples with fertility problems to have children).

energy

The capacity for doing work. Energy cannot be created or destroyed but it can exist in different forms that can be transformed one into the other. For example, **potential energy** (PE) is energy stored in an object as a result of its position, shape or state. An an object raised to a height above the Earth's surface has gravitational PE. As it falls its PE is transformed into **kinetic energy**, the energy of movement.

Although energy is never lost, after a number of conversions it tends to finish up as the kinetic energy of random motion of **molecules** (**heat**) at relatively low **temperatures**. This is 'degraded' energy that is difficult to convert back to other forms.

entropy

In **thermodynamics**, the state of disorder of a system at the atomic, ionic, or molecular level is defined as its entropy. The greater the disorder, the higher the entropy. Thus the fast-moving disordered **molecules** of water **vapour** have higher entropy than those of more ordered **liquid** water, which in turn have more entropy than the molecules in **solid** crystalline ice. In a closed system undergoing change, entropy is a measure of the amount of **energy** unavailable for useful work.

> **ABSOLUTE ZERO = ZERO ENTROPY**
>
> At **absolute zero** (−273.15°C/ −459.67°F/0 K), when all molecular motion ceases and order is assumed to be complete, entropy is zero.

environment

In common usage, 'the environment' often means the total global environment, without reference to any particular organism. Ecologically it is all of the conditions affecting a particular organism, including its physical surroundings, climate, and influences of other living organisms.

See also: *habitat*.

enzyme

A biological **catalyst** produced in **cells**, and capable of regulating the speed of one of the chemical reactions necessary for life. Enzymes are large, complex **proteins**, and are highly specific, each chemical **reaction** requiring its own particular enzyme.

The enzyme's specificity arises from its active site, an area with a shape corresponding to part of the molecule with which it reacts (the substrate). The enzyme and the substrate slot together forming an enzyme–substrate complex that allows the reaction to take place, after which the enzyme falls away unaltered. Temperatures above 60°C/140°F damage (denature) the intricate structure of enzymes, causing reactions to cease. Excessive acidity or alkalinity also denatures enzymes.

PRACTICAL ENZYMES

Enzymes have many medical and industrial uses, from washing powders to **drug** production, and as research tools in molecular biology. They can be extracted from **bacteria** and moulds, and **genetic engineering** now makes it possible to tailor an enzyme for a specific purpose.

equilibrium

An unchanging condition in which the forces and reactions in a system remain in a state of balance so that there is no overall change. In a static equilibrium, such as an object resting on the floor, all the forces acting on it are balanced and there is no motion. In a dynamic equilibrium a steady state is maintained by constant, though opposing, changes. For example, in a sealed bottle half-full of water, the water level stays constant as a result of molecules evaporating from the surface and condensing on to it at the same rate.

ethology

The study of animal behaviour in its natural setting. Ethology is concerned with the causal mechanisms (both the stimuli that elicit behaviour and the physiological mechanisms controlling it), as well as the development of behaviour, its function, and its evolutionary history. Ethologists believe that the significance of an animal's behaviour can be understood only in its natural context, and emphasize the importance of field studies and an

eugenics

The study of ways in which the physical and mental characteristics of the human race may be improved. The term was coined by the English scientist Francis Galton in 1883, and the concept was originally developed in the late 19th century with a view to improving human intelligence and behaviour. Modern eugenics is concerned mainly with the elimination of **genetic disease**.

The eugenic principle was abused by the Nazi Party in Germany during the 1930s and early 1940s to justify the attempted extermination of entire social and ethnic groups and the establishment of selective breeding programmes.

> In 1986 Singapore became the first democratic country to adopt an openly eugenic policy by guaranteeing pay increases to female university graduates when they give birth to a child, while offering grants towards house purchases for nongraduate married women on condition that they are sterilized after the first or second child.

eukaryote

One of the two major groupings into which all **cells** are divided. The cells of eukaryotes possess a clearly defined **nucleus**, bounded by a membrane, within which **DNA** is formed into distinct **chromosomes** during **cell division**. Eukaryotic cells also contain mitochondria, chloroplasts, and other structures (organelles) that, together with a defined nucleus, are lacking in the cells of **prokaryotes**. All cells with the exception of **bacteria** are eukaryotes.

evaporation

The process by which a **liquid** turns to a **vapour** without its **temperature** reaching **boiling point**. A puddle eventually evaporates because, at any time, a proportion of its **molecules** will have enough **kinetic energy** to escape into the atmosphere. The temperature of the liquid tends to fall because the evaporating molecules remove energy from the liquid. This cooling effect may be noticed when wet clothes are worn, or sweat evaporates from the skin. The rate of evaporation rises with increased temperature because as the mean kinetic energy of the liquid's molecules rises, so will the number possessing enough energy to escape.

event horizon
The boundary of a **black hole**. This is the point where the gravitational field surrounding the hole becomes strong enough to prevent the escape of electromagnetic radiation. Anything occuring inside the event horizon can never be observed from outside. Anything crossing the event horizon can never escape.

evolution
The gradual process of change by which **life** has developed by stages from the first single-celled organisms into the present multiplicity of **animal** and **plant** life that inhabits the Earth. The development of the concept of evolution is usually associated with the English naturalist Charles **Darwin** who attributed the main role in evolutionary change to **natural selection** acting on randomly occurring variations within **species** that confer a selective advantage on the individuals displaying them. These **variations** arise from genetic **mutations** that occur spontaneously in all organisms.

experiment
A practical test designed with the intention of supporting or refuting a particular theory or set of theories. Although some experiments may be used merely for gathering more information about a topic that is already well understood, others may be of crucial importance in confirming a new theory or in undermining long-held beliefs. The manner in which experiments are performed, and the relation between the design of an experiment and its value, are therefore of central importance. In general, an experiment is of most value when the factors that might affect the results (variables) are carefully controlled; for this reason most experiments take place in a well-managed environment such as a laboratory or clinic.

extinction
The complete disappearance of a **species**. Extinctions occur when an organism becomes unfit for survival in its natural **habitat**, usually to be replaced by another, better-suited organism

Mass extinctions
Several times in the past large

OVERKILL

About 10,000 years ago many giant mammal species died out. This is known as the 'Pleistocene overkill' because their disappearance was probably hastened by the hunting activities of prehistoric humans.

numbers of species have become extinct virtually simultaneously. The greatest mass extinction occurred about 250 million years ago when up to 96% of all living species became extinct. The best known mass extinction is that of the **dinosaurs**, other large **reptiles**, and various marine **invertebrates** about 65 million years ago. This event has been attributed to catastrophic environmental changes following a meteor impact or unusually prolonged volcanic eruptions.

> **SPECIES LOSS**
>
> The loss of species due to deforestation alone may be 4,000 to 6,000 species a year. Overall, the rate could be as high as one species an hour.

eye

The organ of vision which responds to light. Eyes vary in complexity from the simple structures found in invertebrates to the more complex compound eye of insects. In the human eye, the light is focused by the combined action of the curved cornea, the internal fluids, and the lens. Among molluscs, cephalopods have complex eyes similar to those of vertebrates.

The mantis shrimp's eyes contain ten colour pigments with which to perceive colour; some flies and fishes have five, while the human eye has only three.

F

fallout
The harmful radioactive material released into the atmosphere as the result of a nuclear explosion that subsequently descends to the surface of the Earth. Such material can enter the **food chain**, cause radiation sickness, and last for hundreds of thousands of years (see **half-life**).

Faraday, Michael (1791–1867)
English chemist and physicist who discovered **electromagnetism** and the induction of **electric currents**, and made the first dynamo, the first electric motor, and the first transformer. He also pointed out that the energy of a magnet is in the field around it and not in the magnet itself, extending this basic conception of field theory to electrical and gravitational systems.

Laws of electrolysis
Faraday's laws of **electrolysis** established the link between **electricity** and chemical affinity, one of the most fundamental concepts in science. It was Faraday who coined the terms **anode**, **cathode**, cation, anion, **electrode**, and electrolyte. Faraday demonstrated that the **ions** are discharged at each electrode according to the following rules:

Faraday *Michael Faraday, English chemist and physicist.*

- the quantity of a substance produced is proportional to the amount of electricity passed
- the relative quantities of different substances produced by the same amount of electricity are proportional to their equivalent weights (that is, the relative atomic **mass** divided by the oxidation state or **valency**).

FARADAY'S LIFE

1791 Faraday is born in Newington, Surrey, UK, on 22 September.
1805 Having only received rudimentary education, he is apprenticed to a book binder in London, where he begins to read voraciously.
1810 He is introduced to the City Philosophical Society and receives a basic grounding in science.
1813 Faraday becomes laboratory assistant to Humphry Davy at the Royal Institution.
1831 He discovers electromagnetic induction and invents the first electric generator.
1833 He succeeds Davy as professor of chemistry.
1846 Faraday theorizes on the nature of light and develops a form of the electromagnetic theory of light, which is later developed by James Clerk Maxwell.
1862 He resigns from the Royal Institution and retires to an apartment provided by Queen Victoria.
1867 Faraday dies on the 25 August.

fat

In many **animals** and **plants**, excess **carbohydrates** and **proteins** are converted into **fats** for storage. **Mammals** and other **vertebrates** store fats in specialized connective tissues (adipose tissues), which not only act as energy reserves but also insulate the body and cushion its organs. Fats are essential constituents of food for many animals, with a calorific value twice that of carbohydrates. Fats are a mixture of organic **compounds** called lipids. More specifically, the term refers to a lipid mixture that is **solid** at room temperature (20°C); lipid mixtures that are **liquid** at room temperature are called oils.

Five fat facts

- fat is a source of energy (9 kcal/g)
- it makes the diet palatable
- it provides basic building blocks for cell structure
- it provides essential fatty acids (linoleic and linolenic)
- it acts as a carrier for fat-soluble vitamins (A, D, E, and K).

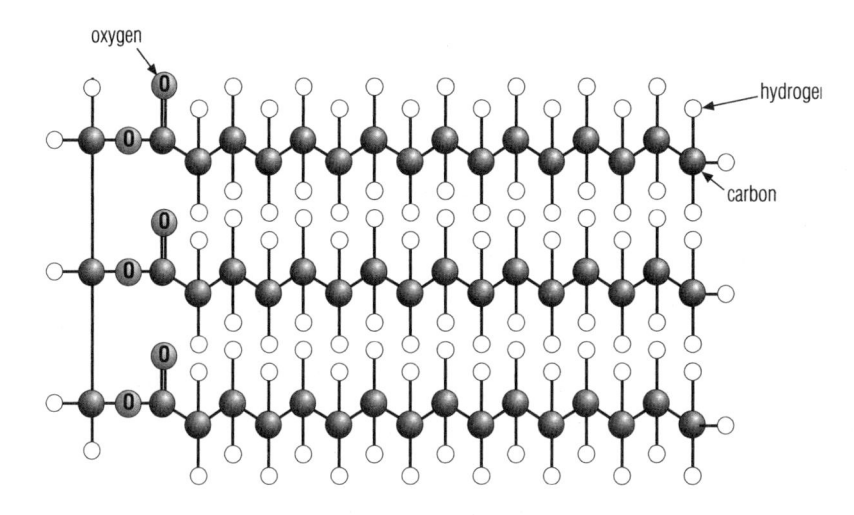

fat *The molecular structure of typical fat. The molecule consists of three fatty acid molecules linked to a molecule of glycerol.*

Eating too much fat, especially fat of animal origin, has been linked with heart disease in humans. Products high in monounsaturated or **polyunsaturated** fats are thought to be less likely to contribute to cardiovascular disease.

feather

The defining feature of **birds**. Feathers are a rigid outgrowth of the outer layer of the skin of birds and are made of the **protein** keratin. Feathers provide insulation and facilitate flight. The colouring of feathers is often important in **camouflage** or in courtship and other displays. Feathers are normally replaced at least once a year.

FEATHER FACTS

A whistling swan has over 25,000 contour feathers (feathers that give the body its streamlined shape), whereas a ruby-throated hummingbird has less than 950.

fermentation

The breakdown of sugars by **anaerobic** respiration. **Bacteria** and yeasts use this method of **respiration** without **oxygen**. Fermentation processes have long been utilized in baking bread, making beer and wine, and producing cheese, yoghurt, soy sauce, and many other foodstuffs. In baking and brewing, yeasts ferment sugars to produce ethanol and carbon dioxide; the latter makes bread rise and puts bubbles into beers and champagne.

fertilization

The fusion of male and female gametes (sex cells, often called egg and sperm) to produce a zygote (a fertilized egg). As each gamete contains only half the correct number of **chromosomes** fertilization results in a **cell** with its full complement of chromosomes, half contributed by each parent. A **plant** self-fertilizes when both the male and female gametes come from the same plant; in cross-fertilization they come from different plants. Self-fertilization rarely occurs in animals; usually even **hermaphrodite** animals (those with both male and female reproductive organs) cross-fertilize each other.

- In terrestrial **insects**, **mammals**, **reptiles**, and **birds**, fertilization occurs within the female's body.
- In the majority of **fishes** and **amphibians**, and most aquatic **invertebrates**, fertilization occurs externally, when both sexes release their gametes into the water.
- In most **fungi**, gametes are not released, but the hyphae of the two parents grow towards each other and fuse to achieve fertilization.
- In higher plants, **pollination** precedes fertilization.

Feynman, Richard Phillips (1918–1988)

US physicist whose work laid the foundations of quantum electrodynamics. For his work on the theory of **radiation** he shared the Nobel Prize for Physics in 1965 with Julian Schwinger (1918–1994) and Sin-Itiro Tomonaga (1906–1979). Feynman contributed to many aspects of **particle physics**, including **quark** theory and the nature of the weak nuclear force (see **fundamental forces**). For his work on quantum electrodynamics, he developed a simple and elegant system of diagrams to represent interactions between particles and how they moved from one **space-time** point to another. He had rules for calculating the probability associated with each diagram, now known as Feynman diagrams.

Feynman's other major discoveries are the theory of superfluidity (frictionless flow) in liquid helium, developed in the early 1950s; his work on the weak interaction (with US physicist Murray Gell-Mann) and the strong force; and his prediction that the **proton** and **neutron** are not elementary particles. Both particles are now known to be composed of quarks. *The Feynman Lectures on Physics* (1963) has become a standard work.

> 6 [Physics] is a major part of the true culture of modern times. 9
>
> **Richard Feynman**, epilogue to *The Feynman Lectures on Physics*, vol. 3

filter

In **chemistry** a device for separating **solid** particles from a **liquid** or **gas**. In **physics** a device placed in front of a beam of **radiation** to alter its **frequency** distribution, for example a coloured filter placed in the path of a beam of **light**.

fish

Fish of some form are found in virtually every body of **water** in the world except for the very salty water of the Dead Sea and some of the hot lava springs. Of the 30,000 fish **species**, approximately 2,500 are freshwater. Fish are aquatic **vertebrates** that use gills to obtain **oxygen** from water. There are three main groups: the bony fishes or Osteichthyes (goldfish, cod, tuna); the cartilaginous fishes or Chondrichthyes (sharks, rays); and the jawless fishes or Agnatha (hagfishes, lampreys).

- *Bony fishes:* the majority of living fishes (about 20,000 species). The **skeleton** is bone, movement is controlled by mobile fins, and the body is usually covered with scales. The gills are covered by a single flap. Many have a swim bladder with which the fish adjusts its buoyancy. Most lay eggs, sometimes in vast numbers; some cod can produce as many as 28 million.

- *Cartilaginous fishes:* there are fewer than 600 known species of sharks and rays. The skeleton is cartilage, the mouth is generally beneath the head, the nose is large and sensitive, and there is a series of open gill slits along the neck region. They have no swimbladder and, in order to remain buoyant, must keep swimming. They may lay eggs ('mermaid's purses') or bear live young.

- *Jawless fishes*: these have a body plan like that of some of the earliest vertebrates that existed before true fishes with jaws evolved. They have no true backbone but a notochord. The lamprey attaches itself to the fishes on which it feeds by a suckerlike rasping mouth. Hagfishes are entirely marine, very slimy, and feed on carrion and injured fishes.

BIGGEST AND SMALLEST

The world's largest fish is the 20 m/66 ft long whale shark.
The smallest is the 7.5 mm/0.3 in dwarf pygmy goby.

fission

The splitting of a heavy atomic **nucleus** into two or more major fragments. It is accompanied by the emission of two or three **neutrons** and the release of large amounts of **nuclear energy**. Fission occurs spontaneously in nuclei of uranium-235, the main fuel used in nuclear reactors. However, the process can also be induced by bombarding nuclei with neutrons because a nucleus that has absorbed a neutron becomes unstable and soon splits. The neutrons released spontaneously by the fission of uranium nuclei may therefore initiate a **chain reaction** that must be controlled if it is not to result in a nuclear explosion.

Fleming, Alexander (1881–1955)

Scottish bacteriologist who discovered the first **antibiotic drug**, penicillin, in 1928. In 1922 he had discovered lysozyme, an antibacterial **enzyme** present in saliva, nasal secretions, and tears. While studying this, he found an unusual mould growing on a culture dish, which he isolated and grew into a pure culture; this led to his discovery of penicillin. It came into use in 1941, largely, as a result of the work of Australian-born British bacteriologist Howard Florey (1898–1968) and German-born British biochemist Ernst Chain (1906–1979). In 1945 the three scientists shared the Nobel Prize for Physiology or Medicine

FLEMING'S LIFE

1881   Fleming is born on 6 August in Lochfield, Ayrshire, UK.
1902   He wins a scholarship to study medicine at St Mary's Hospital Medical School in London.

1928	He is appointed professor at St Mary's and lecturer at the Royal College of Surgeons. He makes his major discovery on the antibiotic properties of penicillin.
1939	Florey and Chain, in Oxford, manage to isolate the substance and purify it, making the large-scale production of penicillin possible. Fleming continues to identify organisms that cause infection and shows how cross-infection can occur between patients in hospital wards.
1944	Fleming is knighted.
1946	He becomes Director of the Wright-Fleming Institute, where he works until he retires.
1955	He dies on 11 March, in London.

floating

The state of **equilibrium** in which a body rests on or is suspended in the surface of a fluid (**liquid** or **gas**). According to **Archimedes' principle**, a body wholly or partly immersed in a fluid will be subjected to an upward force, or upthrust, equal in magnitude to the **weight** of the fluid it has displaced.

If the **density** of the body is greater than that of the fluid, then its weight will be greater than the upthrust and it will sink. However, if the body's density is less than that of the fluid, the upthrust will be the greater and the body will be pushed upwards towards the surface. As the body rises above the surface the amount of fluid that it displaces (and therefore the magnitude of the upthrust) decreases. Eventually the upthrust acting on the submerged part of the body will equal the body's weight, equilibrium will be reached, and the body will float.

flower

The reproductive unit of an **angiosperm** or flowering **plant**, typically consisting of four whorls of modified leaves: sepals, petals, stamens, and carpels. These are borne on a central axis or receptacle. The many variations in size, **colour**, number, and arrangement of parts

BLOOMING MARVELLOUS

Flowers range in size from the tiny blooms of duckweeds that are scarcely visible to the naked eye to the gigantic flowers of the Malaysian *Rafflesia*, which can reach over 1 m/3 ft across.

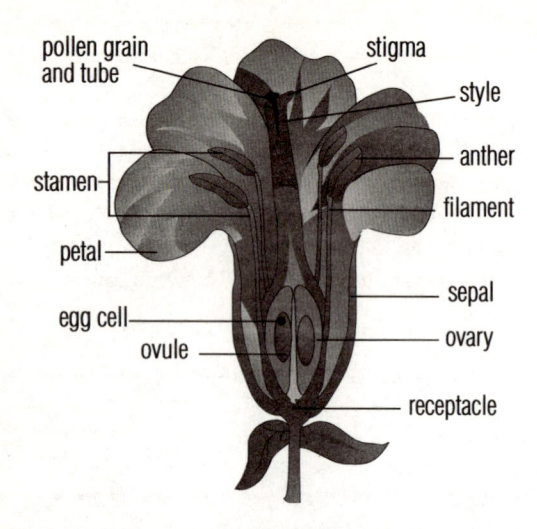

flower *Cross section of a typical flower showing its basic components: sepals, petals, stamens (anthers and filaments), and carpel (ovary and stigma). Flowers vary greatly in the size, shape, colour, and arrangement of these components.*

are closely related to the method of **pollination**. Flowers adapted for wind pollination typically have reduced or absent petals and sepals and long, feathery stigmas that hang outside the flower to trap airborne pollen. In contrast, the petals of **insect**-pollinated flowers are usually conspicuous and brightly coloured.

Flower structure
The sepals and petals form the calyx and corolla respectively and together comprise the perianth which protects the reproductive organs and attracts pollinators. The stamens lie within the corolla, each having a slender stalk, or filament, bearing the pollen-containing anther at the top. Collectively they are known as the androecium (male organs). The inner whorl of the flower comprises the carpels, each usually consisting of an ovary in which are borne the ovules, and a stigma borne at the top of a slender stalk, or style. Collectively the carpels are known as the gynoecium (female organs).

food
Anything consumed by organisms to sustain **life** and health. Humans require the following in their diet:

- *carbohydrates*
- *fats*
- *minerals*
- *proteins*
- *vitamins*
- *water*

Food is needed both for energy, measured in calories or kilojoules, and nutrients, which are converted to body tissues. Some nutrients, such as fat and carbohydrate provide mainly **energy**; other nutrients are important in other ways; for example, dietary fibre is an aid to **metabolism**. Proteins are necessary for building **cell** and tissue structure.

food chain

A simple way of representing the feeding relationships between organisms in a particular **ecosystem**. Each organism depends on the next lowest member of the chain for its food. **Energy** in the form of **food** is transferred along the chain from **autotrophs** (self-feeders), or producers, which are in most ecosystems **plants** and photosynthetic **micro-organisms**, to a series of heterotrophs (other feeders), or consumers. The heterotrophs comprise the **herbivores**, which feed on the producers; **carnivores**, which feed on the herbivores; and decomposers, which break down the dead bodies and waste products of all four groups (including their own), ready for recycling.

In reality, however, the food chain is an oversimplification. The more complex food web shows a greater variety of relationships, but again emphasizes that energy passes from plants to herbivores to carnivores. Energy is lost at every step in the chain and a pyramid of numbers is sometimes used to illustrate this, with fewer and fewer animals being supported by those below them in the chain. (*See illustration on p. 92.*)

force

Any influence that tends to change the state of rest or the uniform motion in a straight line of a body. All moving bodies continue moving with the same **velocity** unless a force is applied to cause an acceleration (a change in velocity over time). A resultant force is a single force acting on a particle or body whose effect results from the combined effects of two or more separate forces. The action of an unbalanced or resultant force results in the acceleration of a body in the direction of action of the force, or it may, if the body is unable to move freely, result in its deformation.

See also: *friction, Galileo, momentum, Newton's laws of motion.*

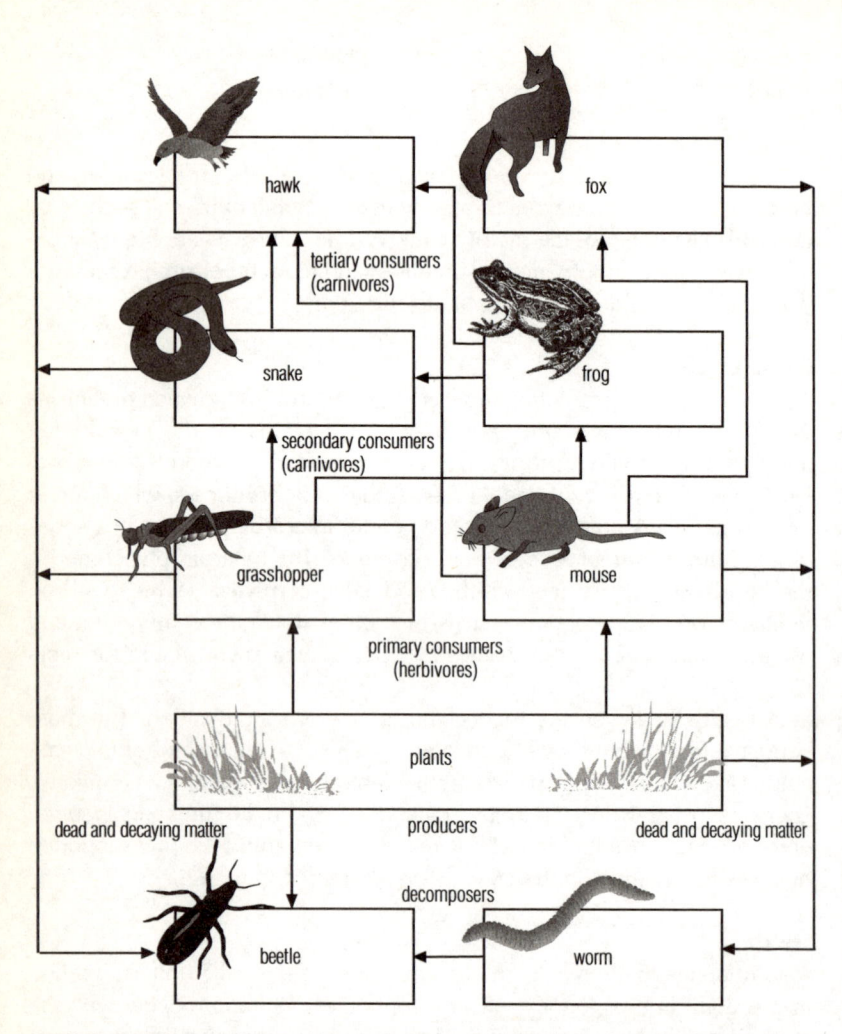

food chain *The complex interrelationships between animals and plants in a food web. A food web shows how different food chains are linked in an ecosystem. Note that the arrows indicate movement of energy through the web. For example, an arrow shows that energy moves from plants to the grasshopper, which eats the plants.*

fossil

A cast, impression, or the actual remains of an **animal** or **plant** preserved in rock. Fossils are created during periods of rock formation, caused by the

In October 1995, a US geologist found the outline of a jellyfish-like animal in sandstone and shale in the Mexican desert. It was estimated to have lived on the seafloor 600 million years ago, and if this is correct it is the oldest known animal fossil.

gradual accumulation of sediment over millions of years at the bottom of the sea bed or an inland lake. Fossils may include footprints, an internal cast, or external impression and can range in size from complete **dinosaur skeletons** to microfossils such as pollen and **bacteria**. A few fossils are preserved intact, as with mammoths fossilized in Siberian ice, or **insects** trapped in tree resin that is today amber. The study of fossils is called palaeontology.

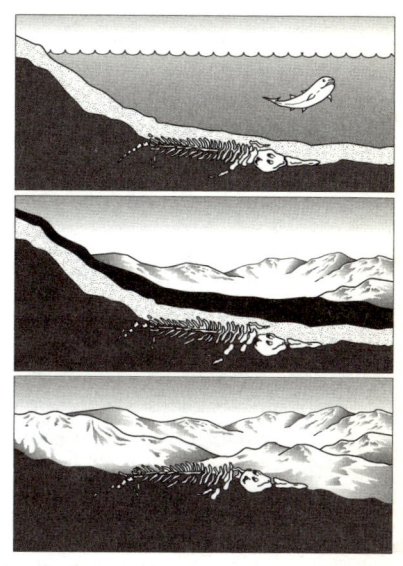

fossil *An animal dies and sinks into mud at the bottom of the sea. Its flesh and bones decay leaving a fossil cast. The layers of mud and sand gradually turn into rock. After millions of years the bed of the sea is raised up and becomes land. The land is eroded by wind and rain and eventually the fossil appears.*

MISSING IN TIME

About 250,000 fossil species have been discovered – a figure that is believed to represent less than 1 in 20,000 of the species that ever lived.

fractal

The irregular shape or surface produced by a procedure of repeated subdivision. The name was coined by the French mathematician Benoit Mandelbrot (1924–). Generated on a computer screen, fractals are used in creating models of geographical or biological processes (for example, the creation of a coastline by erosion or accretion, or the growth of plants). Sets of curves with such discordant properties were developed in the 19th

> **DATA COMPRESSION**
>
> Fractal compression is a method of storing digitally processed picture images as fractals. It uses less than a quarter of the data produced by breaking down images into pixels. The technique was first used commercially in CD-ROM products in 1993.

century in Germany by Georg Cantor and Karl Weierstrass. Fractals are also used for computer art.

Franklin, Benjamin (1706–1790)

US scientist, statesman, writer, printer, and publisher. He proved that lightning is a form of electricity, distinguished between positive and negative **electricity**, and invented the lightning conductor. He was the first US ambassador to France 1776–1785, and negotiated peace with Britain in 1783. He helped to draft the Declaration of Independence and the US Constitution.

Over seven years, beginning in 1746, Franklin executed a remarkable series of **experiments**. He believed that electricity was a single fluid that flows into or out of objects to produce **electric charges**. Franklin also made a fundamental discovery when he realized that the gain and loss of electricity must be balanced – the concept of conservation of charge. Franklin's most famous work of all involved lightning. He suggested that a person could draw a charge from a thundercloud and in 1752 he carried out his famous experiments with kites. By flying a kite in a thunderstorm, he was able to charge up a capacitor and produce sparks from the end of the wet string, which he held with a piece of insulating silk. The lightning conductor used everywhere today owes its origin to these experiments.

Franklin *US scientist Benjamin Franklin's discovery of the ability to draw an electric charge from a thundercloud led to the introduction of lightning conductors as a way of protecting buildings from being struck by lightning.*

> **FRANKLIN'S LIFE**
>
> **1706** Franklin is born in Boston, Massachusetts, USA on 17 January.
> **1730** He starts publishing the *Pennsylvania Gazette*.
> **1732** He achieves great success with the publication of *Poor Richards Almanack*, a collection of articles and advice on a huge range of topics.
> **1737** Franklin becomes Clerk of the Pennsylvania Assembly and postmaster of Philadelphia.
> **1751** He is elected to the Pennsylvania Assembly and starts his most famous research on electricity.
> **1752** Franklin performs his experiments with kites in thunderstorms.
> **1776** He helps to draft the American *Declaration of Independence* and is one of its signatories.
> **1787** He guides the Constitutional Convention in the formulation and ratification of the American Constitution.
> **1789** Franklin dies on 17 June, in Philadelphia.

Franklin, Rosalind Elsie (1920–1958)
English chemist and **X-ray** crystallographer who was the first to recognize the helical shape of **DNA**. Her work, without which the discovery of the structure of DNA would not have been possible, was built into James **Watson** and Francis **Crick's** Nobel prizewinning description of DNA. She died of cancer on 16 April 1958 at the age of 37, four years before Watson, Crick, and Maurice Wilkins were awared the Nobel prize in 1962. The Nobel prize cannot be awarded posthumously.

freezing
The change of a substance from a **liquid** to a **solid** state, as when water becomes ice. For a given substance, freezing occurs at a definite **temperature**, known as the freezing point – the temperature remains at this point until all the liquid is frozen. The amount of heat per unit **mass** that has to be removed to freeze a substance is a constant for any given substance, and is known as the **latent heat** of **fusion**.

frequency
The number of periodic oscillations, **vibrations**, or **waves** occurring per unit of time. The SI unit of frequency is the hertz (Hz), one hertz being equivalent to one cycle per second.

> **HEARING THINGS**
>
> Human beings can hear sounds from objects vibrating in the range 20–15,000 Hz. Ultrasonic frequencies well above 15,000 Hz can be detected by such mammals as bats. Infrasound (low frequency **sound**) can be detected by some animals and birds. Pigeons can detect sounds as low as 0.1 Hz; elephants communicate using sounds as low as 1 Hz.

friction

The force that opposes the motion of one object relative to another with which it is in contact. The coefficient of friction is the ratio of the force required to achieve this relative motion to the force pressing the two bodies together. Friction is greatly reduced by the use of **lubricants** such as oil, grease, and graphite. Air bearings are now used to minimize friction in high-speed rotational machinery. In other instances friction is deliberately increased by making the surfaces rough – for example, brake linings, driving belts, the soles of shoes, and tyres.

fruit

The structure that forms from the ovary of a **flower** following **fertilization**. It consists of the pericarp or fruit wall, which develops from the ovary wall, that encloses one or more **seeds**. Its function is to protect the seeds during their development and to aid in their dispersal. Sometimes parts other than the ovary are incorporated into the fruit structure, resulting in a false fruit or pseudocarp. Fruits may be dehiscent, which open to shed their seeds, or indehiscent, which remain unopened and are dispersed as a single unit. Fruits can be divided botanically into dry and succulent (fleshy). Dry fruits are generally dispersed by wind or water, fleshy fruits are dispersed by animals. Succulent fruits are often edible, sweet, juicy, and colourful. Most fruits are borne by perennial plants.

fundamental forces

The four fundamental interactions believed to be at work in the physical **universe**. There are two long-range forces: **gravity**, which keeps the planets in orbit around the Sun, and acts between all particles that have **mass**; and the electromagnetic force, which stops **solids** from falling apart, and acts between all particles with **electric charge**.

There are two very short-range forces which operate only inside the atomic nucleus: the weak nuclear force, responsible for the reactions that fuel the Sun and for the emission of beta particles from certain nuclei; and the strong nuclear force, which binds together the **protons** and **neutrons** in the nuclei of **atoms**.

By 1971, US physicists Steven Weinberg (1933–) and Sheldon Glashow (1932–), Pakistani physicist Abdus Salam (1926–1996), and others had developed a theory that suggested that the weak and electromagnetic forces were aspects of a single force called the electroweak force; experimental support came from observation at CERN in the 1980s. Physicists are now working on theories to unify all four forces (see **grand unified theory**).

RELATIVE STRENGTHS OF THE FOUR FORCES

strong	1
electromagnetic	10^{-2}
weak	10^{-6}
gravitational	10^{-40}

In other words, gravity is 10,000 billion, billion, billion, billion times weaker than the strong nuclear force!

fungus plural fungi

Any of a unique group of organisms that includes moulds, yeasts, rusts, smuts, mildews, mushrooms, and toadstools. There are around 70,000 species of fungi known to science, though there may be as many as 1.5 million actually in existence. Unlike plants fungi contain no **chlorophyll** and are therefore unable to make their own food by **photosynthesis**; and they reproduce by spores rather than **seeds**.

Fungi are either **parasites**, existing on living **plants** or **animals**, or saprotrophs, living on dead matter. Many of the most serious plant **diseases** are caused by fungi, and several fungi attack humans and animals. Athlete's foot, thrush, and ringworm are fungal diseases. Typically a fungus has a many-branched structure (the mycelium), made up of threadlike chains of cells called hyphae.

LARGEST LIVING THING

In 1992 an individual honey fungus in Washington State, USA, was identified as the world's largest living thing, having an underground network of hyphae covering 600 hectares/1,480 acres. It was estimated to be between 500 and 1,000 years old.

When the fungus is ready to reproduce, the hyphae become closely packed into a solid mass called the fruiting body, which is usually small and inconspicuous but can be very large; mushrooms, toadstools, and bracket fungi are all examples of large fruiting bodies. These carry and distribute the spores.

fusion

The fusing of the nuclei of light **elements**, such as hydrogen, into those of a heavier element, such as helium. The resultant loss in their combined **mass** is converted into **energy**. Stars and thermonuclear weapons are powered by nuclear fusion. Very high **temperatures** and **pressures** are thought to be required in order for fusion to take place. Under these conditions the atomic nuclei can approach each other at high speeds and overcome the mutual repulsion of their positive charges. At very close range, the strong nuclear force comes into play, fusing the particles together to form a larger **nucleus**.

FUTURE POWER

As fusion is accompanied by the release of large amounts of energy, the process might one day be harnessed to form the basis of commercial energy production. So far no successful fusion reactor has been built.

Galileo (1564–1642)

Properly Galileo Galilei, Italian mathematician, astronomer, and physicist. Galileo's work founded the modern scientific method of deducing laws to explain the results of observation and **experiment**. He developed (but did not invent) the astronomical telescope and was the first to see sunspots, the four main satellites of Jupiter, and the appearance of Venus going through phases, thus proving it was orbiting the Sun. His observations supported Polish astronomer Nicolaus Copernicus's

DENIAL DENIED

Forced by the inquisition to deny his belief that the Earth moves around the Sun, Galileo is reputed to have muttered: 'Eppur si muove' ('Yet it does move').

GALILEO'S LIFE

- **1564** Galileo is born in Pisa, Italy on 15 February.
- **1581** He studies medicine at the university which he leaves two years later without taking a degree.
- **1589** He becomes Professor of Mathematics at the University of Pisa.
- **1590** Galileo publishes his first idea on motion in *De Motu/On Motion*.
- **1592** He becomes Professor of Mathematics at Padua.
- **1604** He deduces the law of falling bodies.
- **1610** Galileo publishes his results of his survey of the night sky in *Sidericus nuncius/The Starry Messenger*. He also becomes mathematician and philosopher to the Grand Duke of Tuscany.
- **1632** He publishes *Dialogue Concerning The Two Chief World Systems*.
- **1633** He faces trial on a charge of heresy in April and is sentenced to house arrest.
- **1638** Galileo publishes the summing up of his life's work in *Discources and Mathematical Discoveries Concerning Two New Sciences*.
- **1642** Galileo dies at Arcetri on 8 January.

(1473–1543) theory that the Earth rotated and orbited the Sun.

Galileo discovered that freely falling bodies, heavy or light, have the same, constant acceleration and that this acceleration is due to **gravity**. He also determined that a body moving on a **frictionless** horizontal surface would neither speed up nor slow down. He invented a thermometer, a hydrostatic balance, and a compass, and discovered that the path of a projectile is a parabola.

Galvani, Luigi (1737–1798)

Italian physiologist who discovered galvanic, or voltaic, **electricity** when investigating the contractions produced in the muscles of dead frogs by contact with pairs of different **metals**. His work led quickly to Alessandro **Volta's** invention of the electrical **cell**, and later to an understanding of how nerves control muscles. In 1786 Galvani noticed that touching a frog with a metal instrument during a thunderstorm made the frog twitch. He concluded that electricity was causing the contraction and postulated (incorrectly) that it came from the animal's muscle and nerve tissues. By 1800, Volta had proved that Galvani had been wrong and that the source of the electricity in his experiments had been two different metals (acting as **electrodes**) and the animal's body fluids (acting as an electrolyte). Nevertheless, for many years current electricity was called Galvanic electricity.

gamma radiation

Very high-frequency **electromagnetic** radiation, similar in nature to **X-rays** but of shorter wavelength, emitted by the nuclei of **radioactive** substances during decay or by the interactions of high-energy **electrons** with matter. Cosmic gamma rays have been identified as coming from pulsars, radio galaxies, and quasars, although they cannot penetrate the Earth's atmosphere.

Gamma rays are stopped only by direct collision with the **nucleus** of an **atom** and are therefore very penetrating; they can, however, be stopped by about 4 cm/1.5 in of lead or by a very thick concrete shield. They are less ionizing in their effect than alpha and beta particles, but are dangerous nevertheless because they can penetrate deeply into body tissues such as bone marrow. They are not deflected by either magnetic or electric fields.

gas

A form of **matter**, such as air, in which the **molecules** move randomly in otherwise empty space, filling any size or shape of container into which the gas is put. Gases can be liquefied by cooling, which lowers the speed of the molecules and enables attractive forces between them to bind them together.

> **IN A PUFF OF AIR**
>
> A sugar-lump sized cube of air at room temperature contains 30 trillion molecules moving at an average speed of 500 metres per second (1,800 kph/1,200 mph).

gene

The unit of inheritance. In higher organisms, genes are located on the **chromosomes**. A gene occurs at a particular point, or locus, on a particular chromosome and may have several variants, or **alleles**, each specifying a particular form of the inherited characteristic – for example, the alleles for blue or brown eyes. In the 1940s, it was established that a gene could be identified with a particular length of **DNA**, which coded for a complete protein **molecule,** leading to the 'one gene, one **protein**' principle. Later it was realized that some proteins are made up of several polypeptide chains, each with a separate gene, so this principle was modified to 'one gene, one polypeptide'. However, the fundamental idea remains the same, that genes produce their visible effects simply by coding for proteins. Genes undergo **mutation** and recombination to produce the **variation** on which **natural selection** operates.

gene bank

A collection of **seeds** or other forms of genetic material, such as tubers, spores, bacterial or yeast cultures, live animals and plants, frozen sperm and eggs, or frozen embryos. These are stored for possible future use in agriculture, plant and animal breeding, or in medicine, **genetic engineering**, or the restocking of wild **habitats** where **species** have become extinct. Gene banks will be increasingly used as the rate of **extinction** increases, depleting the Earth's **biodiversity**.

gene therapy

A medical technique for curing or alleviating inherited **diseases** or defects, certain infections, and several kinds of **cancer**, in which a defective **gene** is replaced by a properly functioning one. Cystic fibrosis is the commonest inherited disorder and the one most keenly targeted by gene therapists.

Gene therapy is not the final answer to inherited disease; it may cure the patient but it cannot prevent him or her from passing on the genetic defect

ASHANTHI DE SILVA

The first human being to undergo approved gene therapy, in 1990, was Ashanthi de Silva, then four years old. She had a rare, inherited disorder called ADA deficiency. Children with this condition are vulnerable to a host of common childhood diseases, some of which could prove fatal. Such children are nursed in a germ-free environment. Researchers at the National Institutes of Health in the USA removed white blood cells from her body and infected them with a virus carrying a working copy of the ADA gene. They were then injected back into her body.

to any children, unless the treatment is carried out on the germ (sex) cells. This is a highly controversial area, however, because of the risks of making changes to the human **genome** that would be passed on to succeeding generations.

genetic code

The way in which instructions for building **proteins**, the basic structural **molecules** of living organisms, are 'written' in the genetic material, **DNA**. This relationship between the sequence of bases (the subunits in a DNA molecule) and the sequence of **amino acids** (the subunits of a protein molecule) is the basis of heredity. The code employs codons of three bases each; it is the same in almost all organisms, except for a few minor differences recently discovered in some protozoa.

genetic engineering

The deliberate manipulation of genetic material by biochemical techniques. It is often achieved by the introduction of new **DNA**, usually carried by means of a **virus** or bacterial plasmid. This can be for pure research, **gene therapy**, or to breed functionally specific plants, **animals**, or **bacteria**. For example, plants grown for food could be given the ability to fix nitrogen, found in some bacteria, and so reduce the need for expensive fertilizers.

Organisms to which a foreign **gene** has been added are described as transgenic. By the beginning of 1995 more than 60 plant species had been genetically engineered, nearly 3,000 transgenic crops had been field-tested, and in 1999, a 4-year trial large-scale was begun in Britain to gauge the impact of genetically modified crops on surrounding natural habitats.

GENETIC CONTROVERSY

The first genetically engineered food went on sale in 1994; the 'Flavr Savr' tomato, produced by the US biotechnology company Calgene, was available in California and Chicago. There are concerns for the environmental consequences of genetically modified (GM) crops, for example, in 1999 US ecologists found evidence that maize modified to contain insecticidal *Bt* genes from the soil bacterium *Bacillus thuringiensis* may be harmful to the monarch butterfly caterpillar.

genetic fingerprinting or genetic profiling

A technique developed in the UK by English geneticist Professor Alec Jeffreys (1950–), and now allowed as a means of legal identification. Certain regions of **DNA**, known as hypervariable regions, are unique to individuals. Like conventional fingerprinting, genetic profiling can accurately distinguish humans from one another, with the exception of identical siblings from multiple births. Genetic fingerprinting involves isolating DNA from **cells**, then comparing and contrasting the sequences of component chemicals between individuals. Genetic fingerprinting is used in paternity testing, forensic medicine, and inbreeding studies.

An individual's DNA can be ascertained from a sample of skin, hair, or semen. It can be determined from as little material as a single cell.

genetics

The branch of **biology** concerned with the study of **heredity** and **variation**; it attempts to explain how the characteristics of living organisms are passed on from one generation to the next. Modern geneticists investigate the structure, function, and transmission of **genes**. The science of genetics is based on the work of Austrian biologist Gregor **Mendel**.

FIRST GENE MAP

In 1997 scientists completed a genome map for *Saccharomyces cerevisiae*, common brewer's yeast.

genome

The full complement of **genes** carried by a single set of **chromosomes**. The term may be applied to the genetic information carried by an individual or to the range of genes found in a given **species**. The human genome is made up of approximately 100,000 genes (though there may be as many as 140,000 acording to a 1999 estimate deriving from the **Human Genome Project**).

genotype

The particular set of **alleles** (variants of **genes**) possessed by a given organism. The term is usually used in conjunction with phenotype, which is the product of the genotype and all environmental effects.

geology

The science of the Earth, its origin, composition, structure, and history. Geology is regarded as part of earth science, a more widely embracing subject that brings in meteorology, oceanography, geophysics, and geochemistry.

Branches of geology

- mineralogy (the minerals of Earth)
- petrology (rocks)
- stratigraphy (the deposition of successive beds of sedimentary rocks)
- palaeontology (fossils)
- tectonics (the deformation and movement of the Earth's crust).

geothermal energy

The **energy** extracted for heating and electricity generation from natural steam, hot water, or hot dry rocks in the Earth's crust. Water is pumped down through an injection well where it passes through joints in the hot rocks. It rises to the surface through a recovery well and may be converted to steam or run through a heat exchanger. Dry steam may be directed through turbines to produce **electricity**. It is an important source of energy in volcanically active areas such as Iceland and New Zealand.

germination

The initial stages of growth in a **seed**, spore, or pollen grain. Seeds germinate when they are exposed to favourable external conditions of moisture, **light**, and **temperature**, and when any factors causing **dormancy** have been removed. The process begins with the uptake of water by the **seed**. The

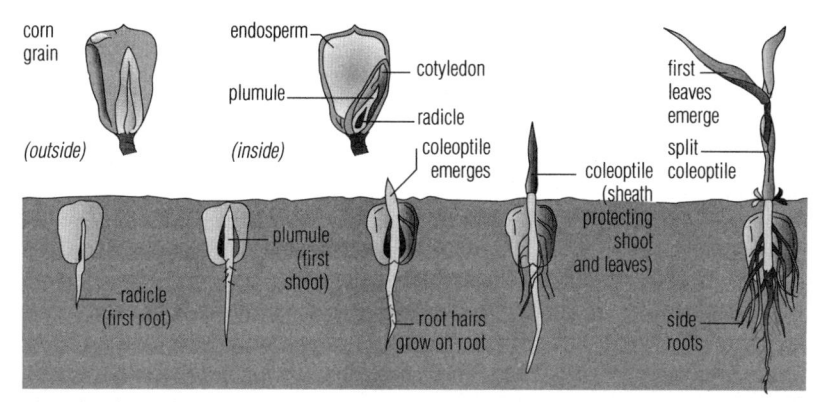

germination *The germination of a corn grain. The plumule and radicle emerge from the seed coat and begin to grow into a new plant. The coleoptile protects the emerging bud and the first leaves.*

embryonic root, or radicle, is normally the first organ to emerge, followed by the embryonic shoot, or plumule. Food reserves, either within the endosperm (the nutritive tissue within the seed) or from the cotyledons (seed leaves), are broken down to nourish the rapidly growing seedling. Germination is considered to have ended with the production of the first true leaves.

grand unified theory

A sought-for theory that would combine the **standard model** of **particle physics**, which explains the **fundamental forces** of **electromagnetism** and the strong and weak nuclear forces, with general **relativity**, which deals with **gravity** and **space-time.**

gravity

The force of attraction that arises between objects by virtue of their masses. On Earth, gravity is the force of attraction between any object in the Earth's gravitational field and the Earth itself. It is regarded as one of the four **fundamental forces** of nature. The gravitational force is the weakest of the four forces, but it acts over

> **GRAVITY AND RELATIVITY**
>
> Albert **Einstein's** general theory of **relativity** treats gravitation not as a force but as the result of objects following the curvature of **space-time** caused by the presence of mass.

Newton's law of gravitation

According to **Newton's** law, all objects fall to Earth with the same acceleration, regardless of mass. For an object of **mass** m_1 at a distance r from the centre of the Earth (mass m_2), the gravitational force of attraction:

$$F \text{ equals } \frac{Gm_1m_2}{r^2}$$

where G is the gravitational constant. However, according to Newton's second law of motion, F also equals m_1g, where g is the acceleration due to gravity; therefore

$$g = \frac{Gm_2}{r^2}$$

and is independent of the mass of the object. At the Earth's surface g equals 9.806 metres per second per second.

greenhouse effect

An atmospheric phenomenon by which solar **radiation**, absorbed by the Earth and re-emitted from the surface as infrared radiation, is prevented from escaping to space by various gases in the air. Greenhouse gases trap heat because they readily absorb **infrared** radiation. The result is a rise in the Earth's temperature.

Principal greenhouse gases

- water vapour
- carbon dioxide
- methane
- chlorofluorocarbons (**CFCs**).

Fossil-fuel consumption and forest fires are the principal causes of carbon dioxide build-up; methane is a by product of agriculture (rice, cattle, sheep). The United Nations Environment Programme estimates that by 2025, average world temperatures will have risen by 1.5°C/2.7°F with a consequent rise of 20 cm/7.9 in in sea level. However, predictions about global warming and its possible climatic effects are tentative and often conflict with each other.

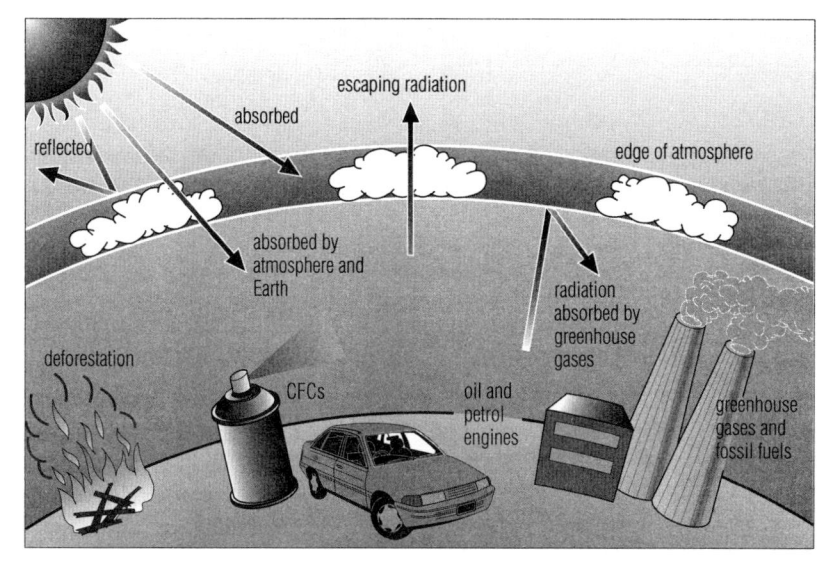

greenhouse effect *The warming effect of the Earth's atmosphere is called the greenhouse effect. Radiation from the Sun enters the atmosphere but is prevented from escaping back into space by gases such as carbon dioxide (produced for example, by the burning of fossil fuels), nitrogen oxides (from car exhausts), and CFCs (from aerosols and refrigerators). As these gases build up in the atmosphere, the Earth's average temperature is expected to rise.*

gymnosperm

Any plant whose **seeds** are exposed, as opposed to the structurally more advanced **angiosperms**, where they are inside an ovary. The group includes conifers and related plants such as cycads and ginkgos, whose seeds develop in cones. **Fossil** gymnosperms have been found in rocks about 350 million years old.

H

habitat
The place in which an organism lives, and which provides for all (or almost all) of its needs. The diversity of habitats found within the Earth's **ecosystem** is enormous, and they are changing all the time. Many can be considered inorganic or physical; for example, the Arctic ice cap, a cave, or a cliff face. Others are more complex; for instance, a woodland or a forest floor. Most habitats provide a home for many species.

Some habitats are so precisely defined that they are called microhabitats, such as the area under a stone where a particular type of insect lives.

half-life
The time taken for half the **atoms** in a sample of radioactive material to decay. In theory, the decay process is never complete and there is always some residual radioactivity. For this reason, the half-life of a radioactive isotope is measured, rather than the total decay time. It may vary from millionths of a second to billions of years.

> **A LONG HALF-LIFE**
>
> The **isotope** of uranium U238 has a half-life of 4.5 billion years, equivalent to the current age of the Earth.

Hawking, Stephen William (1942–)
English physicist whose work in general **relativity** – particularly gravitational field theory – led to a search for a **quantum theory** of **gravity** to explain **black holes** and the **Big Bang** singularities that classical relativity theory does not adequately explain. His book *A Brief History of Time* (1988) gives a popular account of **cosmology** and became an international bestseller.

Hawking's objective of producing an overall synthesis of quantum mechanics and relativity theory began around the time of the publication in 1973 of his seminal book *The Large Scale Structure of **Space-Time***, written with G F R Ellis. His most remarkable hypothesis, published in 1974, was that black holes could in fact emit particles in the form of thermal radiation

– the so-called Hawking **radiation**. The explanation for Hawking radiation relies on the quantum-mechanical concept of 'virtual particles' that exist as particle–antiparticle pairs and are supposed to fill 'empty' space. Hawking suggested that, when such a particle is created near a black hole, one half of the pair might disappear into the black hole, leaving the other half, which could escape to infinity. This would be seen by a distant observer as thermal radiation.

> ❝ If we find why it is that we and the universe exist, it would be the ultimate triumph of human reason – for then we would know the mind of God. ❞
>
> **Stephen Hawking**

heart

The muscular organ that rhythmically contracts to force **blood** around the body of an animal with a circulatory system. Annelid worms and some other **invertebrates** have simple hearts consisting of thickened sections of main blood vessels that pulse regularly. An earthworm has ten such hearts. **Vertebrates** have one heart. A fish heart has two chambers - the thin-walled atrium that expands to receive blood, and the thick-walled ventricle that pumps it out. Amphibians and most reptiles have two atria and one ventricle; birds and mammals have two atria and two ventricles. The beating of the heart is controlled by the autonomic nervous system and an internal control centre or pacemaker, the sinoatrial node.

heat

A form of **energy** possessed by a substance by virtue of the vibrating movement (**kinetic energy**) of its **molecules** or **atoms**. Heat always flows from a region of higher **temperature** (heat intensity) to one of lower temperature. Its effect on a substance may be simply to raise its temperature, or to cause it to expand, melt (if a **solid**), vaporize (if a **liquid**), or increase its **pressure** (if a confined **gas**). Quantities of heat are usually measured in units of energy, such as joules (J) or calories (cal).

Heat energy is transferred by conduction, convection, and radiation.

- *Conduction* is the passing of heat along a solid by collisions and vibrations of its molecules.

- *Convection* is the transmission of heat through a fluid (liquid or gas) in currents – for example, when the air in a room is warmed by a fire or radiator.
- *Radiation* is heat transfer by **infrared** rays. It can pass through a vacuum, travels at the same speed as light, can be **reflected** and **refracted**, and does not affect the medium through which it passes. For example, heat reaches the Earth from the Sun by radiation.

See also: *thermodynamics*.

heavy metal

A metallic **element** of high relative atomic **mass**, such as platinum, gold, and lead. Many heavy metals are poisonous and tend to accumulate and persist in living systems – for example, high levels of mercury (from industrial waste and toxic dumping) accumulate in shellfish and fish, which are in turn eaten by humans. Treatment of heavy-metal poisoning is difficult because available **drugs** are not able to distinguish between the heavy metals that are essential to living cells (zinc, copper) and those that are poisonous.

Heisenberg, Werner Karl (1901–1976)

German physicist who developed **quantum theory** and formulated the uncertainty principle, which places absolute limits on the achievable accuracy of measurement. He was awarded a Nobel prize in 1932 for work he carried out when he was only 24.

THE UNCERTAINTY PRINCIPLE

In 1927, Heisenberg set out his uncertainty principle, which states that there is a theoretical limit to the precision with which a particle's position and **momentum** can be measured. It is impossible to specify precisely both the position and the simultaneous momentum (**mass** multiplied by **velocity**) of a particle. As one value is determined with greater precision, the less exact the other becomes.

Henry, Joseph (1797–1878)

American physicist, inventor of the electromagnetic motor and of a telegraphic apparatus. He also discovered the principle of electromagnetic

induction, roughly at the same time as Michael **Faraday** (although Faraday published first), and the phenomenon of self-induction. He was the Smithsonian Institution's first director, from 1846. His meteorological studies at the Smithsonian led to the founding of the US Weather Bureau.
See also: *electromagnetism.*

herbivore

An animal that feeds on green plants (or **photosynthetic** single-celled organisms) or their products, including seeds, fruit, and nectar. The most numerous type of herbivore is thought to be the zooplankton, tiny **invertebrates** in the surface waters of the oceans that feed on small photosynthetic **algae**. Herbivores are more numerous than other animals because their food is the most abundant. They are the link in the **food chain** between **plants** and **carnivores**.

heredity

The transmission of characteristics from parent to offspring.
See also: *genetics.*

hermaphrodite

An organism that has both male and female sex organs. Hermaphroditism is the norm in some animals such as earthworms and snails, and is common in flowering plants. Cross-fertilization is the rule among hermaphrodites, with the parents functioning as male and female simultaneously, or as one or the other sex at different stages in their development.

Pseudohermaphrodites have the internal sex organs of one sex, but the external appearance of the other. The true sex becomes apparent at adolescence when the internal organs begin to function and release **hormones**.

hibernation

The state of **dormancy** in which certain **animals** spend the winter. It is associated with a dramatic reduction in all **metabolic** processes, including body temperature, breathing, and heart rate. It is a fallacy that hibernating animals sleep throughout the winter.

COLD AND SLOW

- The body temperature of the Arctic ground squirrel falls to below 0°C/32°F during hibernation.
- Hibernating bats may breathe only once every 45 minutes, and can go for up to two hours without taking a breath.

homologous series

Any of a number of series of organic chemicals with similar chemical properties in which members differ by a constant relative molecular mass.

Alkanes (paraffins), alkenes (olefins), and alkynes (acetylenes) form such series in which members differ in mass by 14, 12, and 10 atomic mass units respectively. For example, the alkane homologous series begins with methane (CH_4), ethane (C_2H_6), propane (C_3H_8), butane (C_4H_{10}), and pentane (C_5H_{12}), each member differing from the previous one by a CH_2 group (or 14 atomic mass units).

Hooke, Robert (1635–1703)

English scientist and inventor, originator of Hooke's law, and considered the foremost mechanic of his time. His inventions included a telegraph system, the spirit level, marine barometer, and sea gauge. He studied elasticity, furthered the sciences of mechanics and microscopy (he coined the term **'cell'** in **biology**), invented the hairspring regulator in timepieces, perfected the air pump, and helped improve such scientific instruments as microscopes, telescopes, and barometers. His work on gravitation and in **optics** contributed to the achievements of his contemporary Isaac **Newton**.

> **HOOKE'S LAW**
>
> The deformation of a body is proportional to the magnitude of the deforming force, provided that the body's elastic limit (see **elasticity**) is not exceeded. If the elastic limit is not reached, the body will return to its original size once the force is removed.

hormone

Chemical secretion concerned with control of body functions. Hormones bring about changes in the functions of various organs according to the body's requirements. In addition to the major hormone producing glands

> **MAJOR HORMONE PRODUCING GLANDS**
>
> - thyroid
> - pituitary
> - pancreas
> - testis
> - parathyroid
> - adrenal
> - ovary

there are also hormone-secreting cells in the kidney, liver, gastrointestinal tract, thymus (in the neck), pineal (in the brain), and placenta.

The hypothalamus, which adjoins the pituitary gland at the base of the brain, is a control centre for overall coordination of hormone secretion: the thyroid hormones determine the rate of general body chemistry; the adrenal hormones prepare the organism during stress for 'fight or flight'; and the sexual hormones such as oestrogen and testosterone govern reproductive functions.

Human Genome Project

The research scheme, begun in 1988, to map the complete human genome, a description of the structure of human **DNA**. This is the largest research project ever undertaken in the life sciences.

There are approximately 100,000 different **genes** in the human genome (though a 1999 estimate by a participating US company was as high as 140,000), and one gene may contain more than 2 million nucleotides (the units that make up the DNA molecule).

The programme aims to collect 10–15,000 genetic specimens from 722 ethnic groups whose genetic make-up is to be preserved for future use and study. The knowledge gained is expected to help prevent or treat many crippling and lethal diseases, but there are potential ethical problems associated with knowledge of an individual's genetic make-up, and fears that it will lead to genetic discrimination.

human origins

Analysis of **DNA** in recent human **populations** suggests that *Homo sapiens* (humans) originated about 200,000 years ago in Africa from a single female ancestor, 'Eve'. The oldest known **fossils** of *H. sapiens* also come from Africa, dating from 150,000–100,000 years ago. Bones of the earliest known human ancestor were found in Ethiopia in 1998 and have been dated as 5 million years old. These hominids walked upright and they were either direct ancestors or an offshoot of the line that led to modern humans.

The African apes (gorilla and chimpanzee) have been shown by anatomical and molecular comparisons to be the closest living relatives of humans. Genetic studies indicate that the last common ancestor between chimpanzees and humans lived 5 to 10 million years ago. The oldest known hominids (of the human group), the australopithecines, found in Africa, date from 3.5–4.4 million years ago. The first to use tools came 2 million years

humidity

The quantity of **water** vapour in a given **volume** of the atmosphere (absolute humidity), or the ratio of the amount of water vapour in the atmosphere to the saturation value at the same **temperature** (relative humidity). At dew point the relative humidity is 100% and the air is said to be saturated. Condensation (the conversion of **vapour** to **liquid**) may then occur. Relative humidity is measured by various types of hygrometer.

Huygens, Christiaan (1629–1695)

Dutch mathematical physicist and astronomer. He proposed the **wave** theory of **light**, developed the pendulum clock in 1657, discovered polarization, and observed Saturn's rings. He made important advances in pure mathematics, applied mathematics, and mechanics, which he virtually founded. His work in astronomy was an impressive defence of the Copernican view of the Solar System. Developing the ideas of **Galileo**. He found an accurate experimental value for the distance covered by a falling

> **A TIMELY DEVICE**
>
> In 1657, Huygens developed a clock regulated by a pendulum, an idea that he published and patented. By 1658, major towns in Holland had pendulum tower clocks.

> **HUYGEN'S LIFE**
>
> **1629** Huygens is born on 14 April in The Hague, Netherlands.
> **1645** He is sent to the University of Leiden to study mathematics and law and then spends two years studying law at Breda.
> **1650** He publishes his first study in applied mathematics.
> **1657** Huygens develops a clock regulated by a pendulum.
> **1666** He is invited to live and work at the Bibliotheque Royale in Paris.
> **1673** He publishes *Horologium Oscillatorium*, on the theory of the pendulum.
> **1678** He publishes his most famous work, *Traité de la Lumière*, on his wave theory of light.
> **1695** Huygens dies at The Hague on 8 July.

body in one second. In fact, his gravitational theories successfully deal with several difficult points that **Newton** carefully avoided.

hybrid

The offspring from a cross between individuals of two different **species**, or two inbred lines within a species. In most cases, hybrids between species are infertile and unable to reproduce. In plants, however, doubling of the **chromosomes** can restore the fertility of such hybrids. Hybrids between different genera were believed to be extremely rare but research in the late 1990s shows that hybridization is much more common than traditionally represented. One British evolutionary biologist estimated in 1999 that approximately 10% of animal species and 20% of plant species produced fertile offspring through interspecies mating. Blue whales, for example, hybridize with fin whales and different species of birds of paradise also hybridize. In the wild, a 'hybrid zone' may occur where the ranges of two related species meet.

hydraulics

The field of study concerned with utilizing the properties of **water** and other **liquids**, in particular the way they flow and transmit **pressure**, and with the application of these properties in engineering. It applies the principles of hydrostatics and hydrodynamics. The hydraulic principle of pressurized liquid increasing a force is commonly used on vehicle braking systems, the forging press, and the hydraulic systems of aircraft and excavators.

FIRST USE OF A HYDRAULIC MACHINE

The oldest type of hydraulic machine is the hydraulic press, invented by Joseph Bramah in England in 1795. A hydraulic press consists of two liquid-connected pistons in cylinders, one of narrow bore, one of large bore. A force applied to the narrow piston applies a certain pressure (force per unit area) to the liquid, which is transmitted to the larger piston. Because the area of this piston is larger, the force exerted on it is larger. Thus the original force has been magnified, although the smaller piston must move a great distance to move the larger piston only a little.

hydrocarbon

Any of a class of chemical **compounds** containing only hydrogen and **carbon**. Hydrocarbons are obtained industrially principally from petroleum and coal tar.

hydrogen

A colourless, odourless, gaseous, nonmetallic **element**. Hydrogen, symbol H, is the lightest of all the elements and occurs on Earth chiefly in combination with oxygen as water. Hydrogen is the most abundant element in the universe, where it accounts for 93% of the total number of **atoms** and 76% of the total **mass**. It is a component of most stars, including the Sun, whose heat and light are produced through the nuclear-fusion process that converts hydrogen into helium. When subjected to a pressure 500,000 times greater than that of the Earth's atmosphere, hydrogen becomes a solid with metallic properties, as in one of the inner zones of the planet Jupiter.

Hydrogen's common and industrial uses include the hardening of oils and fats by hydrogenation, the creation of high-temperature flames for welding, and as rocket fuel. It has been proposed as a fuel for road vehicles.

I

ice age
Any of a number of periods during the Earth's history when global temperatures were substantially lowered and ice sheets expanded from the poles towards the equator. The term refers in particular to the ice age that occurred in the Pleistocene epoch, immediately preceding historic times. On the North American continent, glaciers reached as far south as the Great Lakes, and an ice sheet spread over northern Europe as far south as Switzerland. There were several glacial advances separated by interglacial stages during which the ice melted and temperatures were higher than today.

> **SEVERE WEATHER WARNING?**
>
> There is a possibility that the Pleistocene ice age is not yet over. We may be living in an interglacial stage. There might be chilly days ahead...

immunity
The natural protection that organisms have against foreign **microorganisms**, such as **bacteria** and **viruses**, and against cancerous cells (see **cancer**). The cells that provide this protection are the white blood cells, or leucocytes, which make up the immune system. They include neutrophils and macrophages, which can engulf invading organisms and other unwanted material, and natural killer cells that destroy cancerous cells and cells infected by viruses. Some of the most important immune **cells** are the B cells and T cells.

Immune cells coordinate their activities by means of chemical messengers or lymphokines, including the antiviral messenger interferon. The lymph nodes play a major role in organizing the immune response.

inert gas or noble gas
Any of a group of six **elements** (helium, neon, argon, krypton, xenon, and radon), so named because they were originally thought not to enter into

any chemical **reactions**. This is now known to be incorrect: in 1962, xenon was made to combine with fluorine, and since then, compounds of argon, krypton, and radon with fluorine and/or **oxygen** have been described.

The extreme unreactivity of the inert gases is due to the stability of their electronic structure. The outer **electron** shells of inert gas **atoms** are full and they therefore do not need to share or exchange electrons to achieve stability.

Inert gases: electronic structure

Name	Symbol	Atomic number	Electronic arrangement
helium	He	2	2.
neon	Ne	10	2.8.
argon	Ar	18	2.8.8.
krypton	Kr	36	2.8.18.8.
xenon	Xe	54	2.8.18.18.8.
radon	Rn	86	2.8.18.32.18.8.

infrared radiation

Invisible electromagnetic **radiation** of wavelength – between about 0.75 micrometres and 1 millimetre – that lies just beyond the red end of the visible spectrum and just before the shortest microwaves. All bodies above the **absolute zero** of temperature absorb and radiate infrared radiation. Infrared radiation is used in medical photography and treatment, and in industry, astronomy, and criminology. Objects that radiate infrared radiation can be photographed or made visible in the dark on specially sensitized emulsions. This is important for military purposes and in detecting people buried under rubble.

insect

Any of a vast group of small **invertebrate** animals. More than 1 million **species** are known, and several thousand new ones are discovered each year. Insects belong among the **arthropods** and are distributed throughout the world. Many insects hatch out of their eggs as **larvae**, an immature stage bearing no resemblance to the adult, usually in the form of a caterpillar, grub, or maggot. They have to pass

TINIEST INSECT

The world's smallest insect is believed to be a 'fairy fly' wasp in the family Mymaridae, with a wingspan of 0.2 mm/0.008 in.

through further major physical changes (**metamorphosis**) before reaching adulthood. An insect about to go through metamorphosis hides itself or makes a cocoon in which to hide, then rests while the changes take place; at this stage the insect is called a pupa, or a chrysalis if it is a butterfly or moth. When the changes are complete, the adult insect emerges.

The study of insects is called entomology.

Insect characteristics

- hard, segmented bodies composed of chitin
- body divided into three segments:
- head, with a pair of antennae, or feelers
- thorax, where three pairs of jointed legs and usually two pairs of wings are attached
- abdomen, where digestive and reproductive organs are located.

instinct

Behaviour that is presumed to be genetically determined rather than learned, although learning may play a part in its development. Examples include a male robin's tendency to attack other male robins intruding on its **territory** and the tendency of many female **mammals** to care for their offspring. Instincts differ from **reflexes** in that they involve very much more complex actions.

interference

Two or more **waves** can interact and combine with one another to produce resultant waves of larger or smaller amplitude (depending on whether the combining waves are in or out of phase with each other). The result is an **interference** pattern. Interference of white light results in spectral coloured fringes; for example, the iridescent colours of oil films seen on water or soap bubbles. Interference of **sound** waves of similar **frequency** produces the phenomenon of beats, often used by musicians when tuning an instrument. (*See illustration on p. 120.*)

invertebrate

An animal without a backbone. The invertebrates comprise over 95% of existing animal **species** and include sponges, coelenterates, flatworms, nematodes, annelid **worms**, **arthropods**, **molluscs**, echinoderms, and primitive aquatic chordates, such as sea squirts and lancelets.

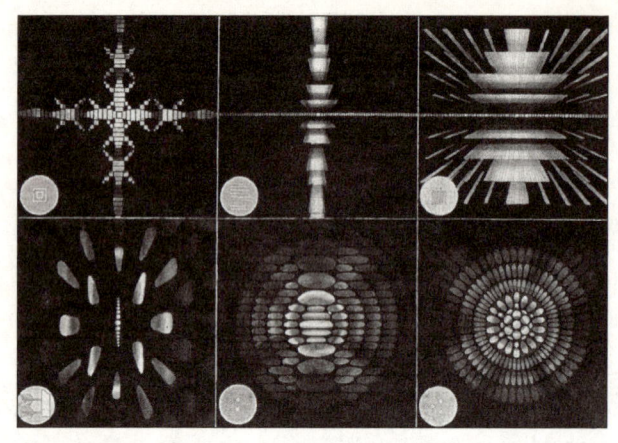

interference *Interference patterns formed by the interference of light waves passed through various apertures.*

ion

An **atom**, or **molecule**, that is either positively charged (cation) or negatively charged (anion), as a result of the loss or gain of **electron**s during chemical **reactions** or exposure to certain forms of **radiation**. In **solution** or in the molten state, ionic **compounds** such as **salts**, **acids**, **alkalis**, and metal oxides conduct **electricity**. These compounds are known as electrolytes.

isotope

One of two or more **atoms** of the same **element** that have the same number of **protons**, but which have a different number of **neutrons**, thus differing in their atomic **mass**. They may be stable or radioactive, naturally occurring, or synthesized. For example, hydrogen has the isotopes ^{2}H (deuterium) and ^{3}H (tritium). The term was coined by English chemist Frederick Soddy, pioneer researcher in atomic disintegration.

J–K

Jenner, Edward (1749–1823)
English physician who pioneered vaccination (see **vaccine**). In Jenner's day, smallpox was a major killer. His discovery in 1796 that inoculation with cowpox gives **immunity** to smallpox was a great medical breakthrough. Jenner observed that people who worked with cattle and contracted the mild disease cowpox from them never subsequently caught smallpox. In 1798 he published his findings that a child inoculated with cowpox, then two months later with smallpox, did not get smallpox. He coined the word 'vaccination' from the Latin word for cowpox, *vaccinia*.

Joule, James Prescott (1818–1889)
English physicist whose work on the relations between electrical, mechanical, and chemical effects led to the discovery of the first law of **thermodynamics**. Joule determined the mechanical equivalent of **heat** (Joule's equivalent) in 1843, and the **SI** unit of **energy**, the joule, is named after him. He also formulated Joule's law, which defines the relation between heat and **electricity.**

kinetic energy
The **energy** of a body resulting from motion. It is contrasted with **potential energy**. The kinetic energy of a moving body is equal to the work that would have to be done in bringing that body to rest, and is dependent upon both the body's **mass** and speed. All **atoms** and **molecules** possess kinetic energy because they are all in some state of motion (see **kinetic theory**). Adding **heat** energy to a substance increases its kinetic energy and hence the movement of its constituent molecules – a change that is reflected as a rise in the **temperature** of that substance.

kinetic theory
The theory describing the physical properties of **matter** in terms of the movement of its component **atoms** or **molecules**. The **temperature** of a substance is dependent on the **velocity** of movement of its constituent particles, increased temperature being accompanied by increased movement. A **gas**

consists of rapidly moving atoms or molecules and, according to kinetic theory, it is their continual impact on the walls of the containing vessel that accounts for the **pressure** of the gas. The slowing of molecular motion as temperature falls, according to kinetic theory, accounts for the physical properties of **liquids** and **solids**, culminating in the concept of no molecular motion at **absolute zero**.

Koch, (Heinrich Hermann) Robert (1843–1910)

German bacteriologist who devised techniques for culturing **bacteria** outside the body, and formulated a set of rules (Koch's postulates) for showing whether or not a bacterium is the cause of a **disease**. His techniques enabled him to identify the bacteria responsible for tuberculosis (1882), cholera (1883), and other diseases. Koch was a great teacher and many of his pupils became outstanding scientists. He was awarded the Nobel Prize for Physiology or Medicine in 1905.

L

larva

The stage between hatching and adulthood in those **species** in which the young have a different appearance and way of life from the adults. Examples include tadpoles (frogs) and caterpillars (butterflies and moths). Larvae are typical of the **invertebrates**, some of which (for example, shrimps) have two or more distinct larval stages. Among **vertebrates**, it is only the **amphibians** and some **fishes** that have a larval stage.

The process whereby the larva changes into another stage, such as a pupa (chrysalis) or adult, is known as **metamorphosis**.

laser

Acronym for light amplification by stimulated emission of radiation, a device for producing a narrow beam of **light**, capable of travelling over vast distances without dispersion. An **atom** in an excited **energy** state has gained enough energy to emit a **photon** of light; this can occur, for example, by collision with another atom or by irradiation with light of suitable wavelength. Normally the atom will emit its photon very quickly and at random, but if a photon of the same wavelength passes while the atom is still in an excited state the atom will emit its photon in phase with the passing photon (stimulated emission). In a laser it is arranged that this process takes place in a manner that causes a rapid build-up of light intensity. The process of providing the atoms with energy is called 'pumping'.

> The Petawatt, the world's most powerful laser, generates 1,200 times as much power as the entire electrical grid of the USA. When focused on platinum, it turns it into gold.

LASER CLOTH-CUTTERS

Carbon dioxide gas lasers, which can produce a beam of over 100 watts in the **infrared**, are used to cut material for suits and dresses in hundreds of thicknesses at a time.

Uses of lasers
- communications (a laser beam can carry much more information than can radio waves)
- cutting
- drilling
- welding
- satellite tracking
- medical and biological research
- surgery.

latent heat
The heat absorbed or released by a substance as it changes state (for example, from **solid** to **liquid**) at constant **temperature** and **pressure**.
See also: *states of matter*.

Lavoisier, Antoine Laurent (1743–1794)
French chemist who proved that combustion needs only a part of the air, which he called **oxygen**, thereby destroying the theory of phlogiston (an imaginary 'fire element' released during combustion). With astronomer and mathematician Pierre de Laplace (1749–1827), he showed in 1783 that water is a compound of oxygen and hydrogen. In this way he established the basic rules of chemical combination. Lavoisier established that organic **compounds** contain **carbon**, hydrogen, and oxygen. From quantitative measurements of the changes during breathing, he showed that carbon dioxide and water are normal products of **respiration**.

Leeuwenhoek, Anton van (1632–1723)
Dutch pioneer of microscopic research. He ground his own lenses, some of which magnified up to 300 times. With these he was able to see individual red blood cells, sperm, and **bacteria**, achievements not repeated for more

---

TINY LENSES

Leeuwenhoek ground more than 400 lenses, which he mounted in various ways in single-lens microscopes. Most of his lenses were very small, some no bigger than a pinhead.

than a century. From 1672 to 1723 he described and illustrated his observations in more than 350 letters to the Royal Society of London. In 1674 he discovered protozoa, which he called 'animalicules', and calculated their sizes. He also studied the structure of the lens in the eye, muscle striations, insect mouthparts, the fine structure of plants, and discovered **parthenogenesis** in aphids. His fame was such that he was visited by several reigning monarchs, including Frederick I of Prussia and Tsar Peter the Great.

lens

A piece of a transparent material, such as glass, that modifies rays of **light**. A lens has two polished surfaces – one concave or convex, and the other plane, concave, or convex. A convex lens brings rays of **light** together; a concave lens makes the rays diverge. Lenses are essential to spectacles, microscopes, telescopes, cameras, and almost all optical instruments.

Aberration
The image formed by a single lens suffers from several defects or aberrations, notably spherical aberration in which an image becomes blurred, and chromatic aberration in which an image in white light tends to have coloured edges. Aberrations are corrected by the use of compound lenses, which are built up from two or more lenses of different refractive index.

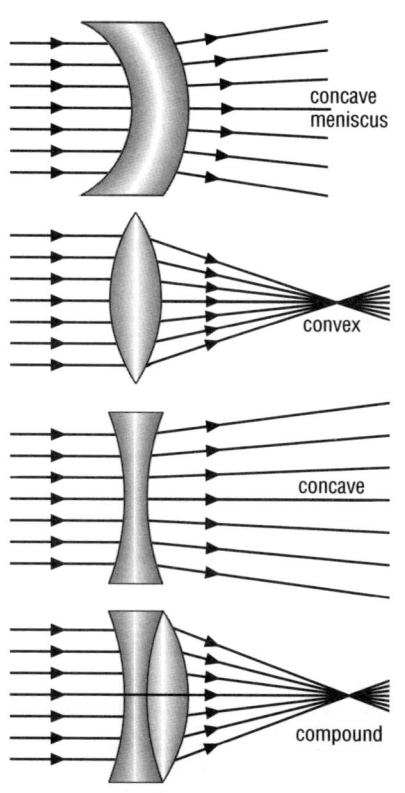

lens *The passage of light through lenses. The concave lens diverges a beam of light from a distant source. The convex and compound lenses focus light from a distant source to a point. The distance between the focus and the lens is called the focal length. The shorter the focus, the more powerful the lens.*

life

Although biologists have a vast knowledge of living things, they find difficulty in defining life and locating the dividing line between living and nonliving things. For example, a **virus** is a lifeless particle until it becomes active inside a living **cell**. Almost all living organisms share certain basic characteristics such as the ability to grow, reproduce, and respond to such stimuli as **light**, **heat**, and **sound**, but not every organism displays all these features, and even inorganic substances may exhibit some of them. Life on Earth may have began about 4 billion years ago when a chemical **reaction** produced the first organic substance. Over time, life has evolved from primitive single-celled organisms to complex multicellular ones. There are now some 10 million different **species** of **plants** and **animals** living on the Earth.

Checklist for life

- reproduction
- growth
- **metabolism**
- movement
- responsiveness
- **adaptation**.

life cycle

The sequence of developmental stages through which members of a given **species** pass. Most **vertebrates** have a simple life cycle consisting of **fertilization** of sex cells or gametes, a period of development as an **embryo**, a period of juvenile growth after hatching or birth, an adulthood including sexual reproduction, and finally death. **Invertebrate** life cycles are generally more complex and may involve completely different appearances and styles of life for **larva** and adult with major reconstitution of the individual's appearance (**metamorphosis**). Some **parasites** have particularly complex life cycles, with different stages taking place in different host organisms.

light

Electromagnetic **waves** that can be detected by the eye. Light, in common with all electromagnetic waves, is considered to exhibit both particle and wave properties. The

---

The **speed of light** (and of all electromagnetic radiation) in a **vacuum** is approximately 300,000 km/ 186,000 miles per second.

emission of light from self-luminous bodies is an atomic phenomenon that has been explained in terms of the **quantum theory**. Certain properties of light, however, are explained only on the hypothesis that light is propagated as electromagnetic waves.

> **WAVELENGTHS**
>
> Light wavelengths go from about 400 nanometres in the extreme violet to about 770 nanometres in the extreme red.

Thus the quantum theory accounts for the photoelectric effect (see **Einstein**), while the electromagnetic wave theory accounts for the **interference** of light. The fundamental particle, or quantum, of light is called the **photon**.

Linnaeus, Carolus (1707–1778) (Latinized form of Carl von Linné) Swedish naturalist and physician whose botanical work *Systema naturae*, 1735, contained his system for classifying plants into groups depending on shared characteristics (such as the number of stamens in **flowers**). This provided biologists with a much-needed framework for identification. He also devised the concise and precise binomial system for naming plants and animals, using one Latin (or Latinized) word to represent the genus and a second to distinguish the **species**. Linnaeus's system of nomenclature was introduced in *Species plantarum*, 1753

See also: *classification*.

> **WHAT'S IN A NAME?**
>
> The Latin name of the daisy is *Bellis perennis*.
>
> - *Bellis* is the name of the genus to which the plant belongs
> - *perennis* distinguishes the species from others of the same genus.

liquid

The state of matter between a **solid** and a **gas**. A liquid forms a level surface and assumes the shape of its container. Its **atoms** do not occupy fixed positions as in a crystalline solid, nor do they have freedom of movement as in a gas. Unlike a gas, a liquid is difficult to compress since **pressure** applied at one point is equally transmitted throughout (Pascal's principle). **Hydraulics** makes use of this property.

lubricant

Substance used between moving surfaces to reduce **friction**. **Carbon**-based (organic) lubricants, commonly called grease and oil, are recovered from petroleum distillation. Extensive research has been carried out on chemical additives to lubricants to reduce corrosive wear, prevent the accumulation of 'cold sludge' (often the result of stop-start driving in city traffic jams), keep pace with the higher working temperatures of aviation gas turbines, or provide **radiation**-resistant greases for nuclear power plants.

luminescence

The emission of **light** from a body when its **atoms** are excited by means other than raising its **temperature**. Short-lived luminescence is called fluorescence; longer-lived luminescence is called phosphorescence. When exposed to an external source of **energy**, the outer **electrons** in atoms of a luminescent substance absorb energy and 'jump' to a higher energy level. When these electrons 'jump' back to their former level they emit their excess energy as light.

Some living organisms produce **bioluminescence**.

WAYS TO GLOW

Many different exciting mechanisms are possible:
- electromagnetic radiation
    visible light
    ultraviolet rays
    X-rays
- electron bombardment
- chemical reactions
- **friction**
- radioactivity (see: **radiation**).

M

magnetism

A group of phenomena associated with magnetic fields. Magnetic fields are produced by moving charged particles: in electromagnets, **electrons** flow through a coil of wire connected to a **battery**; in permanent magnets, spinning electrons within the **atoms** generate the field.

Substances differ in the extent to which they can be magnetized by an external field (their susceptibility). Materials that can be strongly magnetized, such as iron, cobalt, and nickel, are said to be ferromagnetic. Areas called domains form in the magnetized material in which atoms, weakly

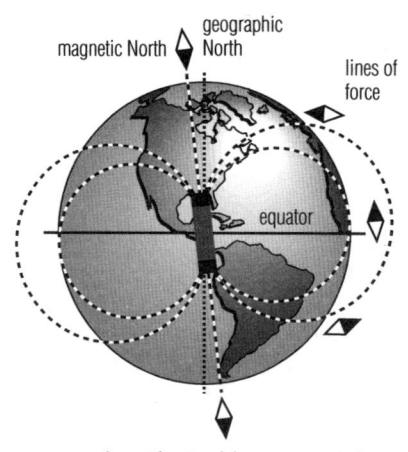

magnetism *The Earth's magnetic field is similar to that of a bar magnet with poles near, but not exactly at, the geographic poles. Compass needles align themselves with the magnetic field, which is horizontal near the equator and vertical at the magnetic poles.*

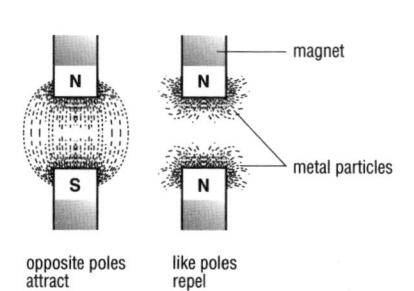

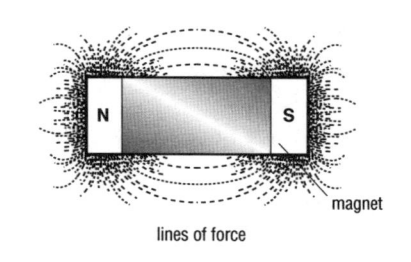

magnetism *Magnets exert invisible lines of force through their poles. If iron filings are sprinkled on to some paper on top of some magnets, their magnetic fields can be 'seen'. The iron filings show that the lines of force are concentrated at the magnets' poles and that like poles repel each other and unlike poles attract.*

magnetic because of their spinning electrons, align to form areas of strong magnetism. Most other materials are paramagnetic, being only weakly pulled towards a strong magnet. This is because their atoms have a low level of magnetism and do not form domains.

> **POLE TO POLE**
>
> When a magnet is broken in two, the result is not two halves, one with a north pole, the other with a south pole, but two new magnets, each with a north and south pole. However often this process is repeated the same result is obtained – every magnet has two poles.

mammal

Any of a large group of warm-blooded **vertebrate** animals characterized by mammary glands in the female that are used for suckling the young.

> **LITTLE AND LARGE**
>
> The smallest shrew weighs only 2 g/0.07 oz, the largest whale up to 140 tonnes.

Mammal features
- mammary glands (in female)
- hair (very reduced in some species, such as whales)
- a middle ear formed of three small bones (ossicles)
- a lower jaw consisting of two bones only
- seven vertebrae in the neck
- no nucleus in the red blood cells.

Mammals are divided into three groups:
- *Placental mammals,* where the young develop inside the mother's body, in the uterus, receiving nourishment from the blood of the mother via the placenta.
- *Marsupials,* where the young are born at an early stage of development and develop further in a pouch on the mother's body where they are attached to and fed from a nipple.
- *Monotremes,* where the young hatch from an egg outside the mother's body and are then nourished with milk.

mass

The quantity of **matter** in a body as measured by its **resistance** to acceleration, or inertia. A body's mass therefore determines the acceleration produced by a given force acting on it. The greater the mass the greater the force need to produce the required acceleration. In the **SI system**, the base unit of mass is the kilogram. Mass can also be defined in terms of the gravitational force. At a given place, equal masses experience equal gravitational forces, which are known as the **weights** of the bodies. Masses may, therefore, be compared by comparing the weights of bodies at the same place.

matter

Anything that has **mass**. All matter is made up of **atoms**, which in turn are made up of elementary particles; it ordinarily exists in one of three physical states: **solid**, **liquid**, or **gas**, depends on conditions of **temperature** and **pressure**. **Kinetic theory** describes how the state of a material depends on the movement and arrangement of its atoms or **molecules**.

> **CONSERVATION OF MATTER**
>
> In chemical **reactions** matter is conserved, so no matter is lost or gained and the sum of the mass of the reactants will always equal the sum of the end products.

Maxwell, James Clerk (1831–1879)

Scottish physicist whose main achievement was in the understanding of **electromagnetic** waves: Maxwell's equations bring together **electricity**, **magnetism**, and **light** in one set of relations. He studied **gases**, **optics**, and the sensation of **colour**, and his theoretical work in magnetism prepared the way for wireless telegraphy and telephony. In developing the **kinetic theory** of gases, Maxwell gave the final proof that **heat** resides in the motion of **molecules**.

Studying colour vision, Maxwell explained how all colours could be built up from mixtures of the primary colours red, green, and blue. Maxwell confirmed English physicist Thomas **Young's** theory that the eye has three kinds of receptors sensitive to the primary colours, and showed that colour blindness is due to defects in the receptors. In 1861 he produced the first colour photograph to use a three-colour process.

MAXWELL'S LIFE

1831	Maxwell is born in Edinburgh, UK on 13 November.
1854	He becomes Professor of Natural Philosophy at Marischal College, Aberdeen.
1855–56	He publishes his first paper on the electromagnetic theory of light, entitled *On Faraday's Lines of Force*.
1860	He becomes Professor of Natural Philosophy and Astronomy at King's College, London.
1861–62	Maxwell publishes his paper *On the Physical Lines of Force*.
1871	He becomes the first Professor of Experimental Physics at Cambridge University.
1873	He publishes a summary of his work in *Treatise On Electricity and Magnetism*.
1874	Maxwell sets up the Cavendish Laboratory, Cambridge.
1879	Maxwell dies on 5 November at the early age of 48.

melting point

The **temperature** at which a substance melts, or changes from **solid** to **liquid** form. A pure substance under standard conditions of **pressure** (usually one atmosphere) has a definite melting point. If **heat** is supplied to a solid at its melting point, the temperature does not change until the melting process is complete.

See also: *states of matter.*

Mendel, Gregor (1822–1884)

Austrian biologist and founder of **genetics**. His **experiments** with successive generations of peas gave the basis for his theory of inheritance involving dominant and recessive characters. Mendel formulated two laws now recognized as fundamental laws of **heredity**: the law of segregation and the law of independent assortment of characters. Mendel concluded that each parent plant contributes a 'factor' (what we

> Mendel reported his findings in *Experiments with Plant Hybrids* in 1866, but the importance of his work was not recognized at the time. It was not until 1900, when his work was rediscovered by Hugo De Vries and others, that Mendel achieved fame – 16 years after his death.

now know as a **gene**) to its offspring that determines a particular trait and that the pairs of factors in the offspring do not give rise to a blend of traits.

Mendeleyev, Dmitri Ivanovich (1834–1907)

Russian chemist who framed the periodic law in **chemistry** in 1869, which states that the chemical properties of the **elements** depend on their relative atomic masses. This law is the basis of the **periodic table** of the elements, in which the elements are arranged by atomic number and organized by their related groups. Mendeleyev was the first chemist to understand that all elements are related members of a single ordered system. From his table he predicted the properties of elements then unknown, of which three (gallium, scandium, and germanium) were discovered in his lifetime.

metabolism

The sum of all the chemical processes that go on in living organisms enabling them to grow and to function. It involves a constant alternation of building up complex **molecules** (anabolism) and breaking them down (catabolism). For example, green **plants** build up complex organic substances from water, carbon dioxide, and mineral salts (**photosynthesis**); **animals** partially break down complex organic substances, ingested as food, and subsequently resynthesize them for use in their own bodies (see **digestive system**). Within **cells**, complex molecules are broken down by the process of **respiration**. The waste products of metabolism are removed by excretion.

metal

Any of a class of chemical **elements** with specific physical and chemical characteristics. Metallic elements compose about 75% of the 112 elements in the **periodic table** of the elements.

Metal properties
- a sonorous tone when struck
- good conduction of **heat** and **electricity**
- lustrous in appearance
- malleability
- ductility.

The majority of metals are found in nature in a combined form only, as **compounds** or **mineral** ores; about 16 of them also occur in the elemental form, as native metals. Their chemical properties are largely determined by the extent to which their **atoms** can lose one or more **electrons** and form positive **ions** (cations). **Nonmetals** typically form negative ions (anions). All metals except mercury are **solid** at ordinary **temperatures**, and all of them will crystallize under suitable conditions. The chief chemical properties of metals also include their strong affinity for certain nonmetallic elements, for example sulphur and chlorine, with which they form sulphides and chlorides. Metals will, when fused, enter into the forming of **alloys**.

The science and technology of producing metals is called metallurgy.

metamorphosis
The period during the **life cycle** of many **invertebrates**, most **amphibians**, and some **fish**, during which the individual's body changes from one form to another through a major reconstitution of its tissues. For example, tadpoles metamorphose into adult frogs; caterpillars metamorphose into butterflies within a pupa.

microbiology
The study of **micro-organisms**, mostly **viruses** and single-celled organisms such as **bacteria**, protozoa, and yeasts. The practical applications of microbiology are in medicine (since many micro-organisms cause **disease**); in brewing, baking, and other food and beverage processes, where the micro-organisms carry out **fermentation**; and in **genetic engineering**, which is creating increasing interest in the field of microbiology.

micro-organism (or microbe)
A living organism invisible to the naked eye but visible under a microscope. Micro-organisms include **viruses** and single-celled organisms such as **bacteria**, **protists**, yeasts, and some **algae**. The term has no significance in biology as a class of organisms. The study of micro-organisms is known as microbiology.

mimicry
The imitation of one **species** (or group of species) by another. Mimicry is usually based on appearance but calls, songs, scents, and other signals can also be mimicked. In some cases, mimicry is not for protection, but allows the mimic to prey on, or parasitize, the model.

- *Batesian mimicry* (named after English naturalist H W Bates (1825–1892) is where the warning colour of an animal that is poisonous or unpleasant to eat is mimicked; the mimic thus benefits from the fact that predators have learned to avoid the organism being mimicked.

- *Mullerian mimicry* is where two or more equally poisonous or distasteful species have a similar colour pattern, thereby reinforcing the warning each gives to predators.

mineral

A naturally occurring substance with a particular chemical composition and usually a regularly repeating crystalline structure. Either in their perfect crystalline form or otherwise, minerals are the constituents of rocks. In more general usage, a mineral is any substance economically valuable for mining (including coal and oil, despite their organic origins).

Mineral forming processes

- *melting* of pre-existing rock and subsequent crystallization of a mineral to form volcanic rocks

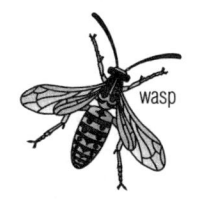

mimicry *Batesian mimicry in which a harmless hoverfly is coloured like an unpleasant wasp in order to confuse a predator. A predator that has tried to eat a wasp will avoid the hoverfly.*

- *weathering* of rocks exposed at the land surface, with subsequent transport and grading by surface waters, ice or wind to form sediments
- *recrystallization* through increasing temperature and pressure with depth to form metamorphic rocks.

mitochondria singular mitochondrion

The membrane-enclosed organelles within **eukaryotic** cells, containing **enzymes** responsible for energy production during **aerobic** respiration. Mitochondria absorb O_2 and glucose and produce energy in the form of **ATP** by breaking down the glucose to CO_2 and H_2O. These rodlike or spherical bodies are thought to be derived from free-living bacteria that, at a very early stage in the history of life, invaded larger cells and took up a symbiotic way of life inside.

molecule

The smallest particles of an **element** or **compound** that can exist independently. Hydrogen **atoms**, for example, do not exist independently under normal conditions. They are bonded in pairs to form hydrogen molecules. A molecule of a compound consists of two or more different atoms bonded together. Each molecule can be represented by a chemical formula, indicating the elements from which it is made and the ratio of each type of atom present. Molecules vary in size and complexity from the hydrogen molecule (H_2) to the large macromolecules of **proteins**. They may be held together by ionic **bonds**, in which the atoms gain or lose **electrons** to form **ions**, or by covalent bonds, where atoms share electrons.

> **MAKING MOLECULES**
>
> A wheel-shaped molecule containing 700 atoms was built by German chemists in 1995. Containing 154 molybdenum atoms surrounded by **oxygen** atoms, it belongs to the class of compounds known as metal clusters.

mollusc

The majority of molluscs are marine **animals**, but some live in fresh water, and a few live on dry land. The molluscs are a group of **invertebrate** animals (Phylum Mollusca), most of which have a body divided into three parts: a head, a central mass containing the main organs, and a foot for movement. The body is soft, without limbs (except for the cephalopods such as octopuses and squid which have arms to capture their prey) and cold-blooded. There is no internal **skeleton**, but many **species** have a hard shell covering the body.

Mollusca

- bivalves (clams, mussels, and oysters)
- gastropods (snails and slugs)
- cephalopods (cuttlefish, squids, and octopuses).

momentum

The momentum of a body does not change unless a resultant or unbalanced force acts on that body to alter its **velocity** (see **Newton's laws of motion**). If the **mass** of a body is m kilograms and its velocity is v then its momentum is given by momentum = mv.

Its unit is the kilogram metre-per-second (kg ms^{-1}) or the newton second. According to Newton's second law of motion, the magnitude of a resultant force F equals the rate of change of momentum brought about by its action, or:

$$F = \frac{(mv - mu)}{t}$$

where mu is the initial momentum of the body, mv is its final momentum, and t is the time in seconds over which the force acts.

The change in momentum, or impulse, produced can therefore be expressed as:

$$\text{impulse} = mv - mu = Ft$$

Conservation of momentum
This is one of the fundamental concepts of classical **physics**. It states that the total momentum of all bodies in a closed system is constant and unaffected by processes occurring within the system.

The angular momentum of an orbiting or rotating body of mass m travelling at a velocity v in a circular orbit of radius R is expressed as mvR. Angular momentum is conserved, and should any of the values alter (such as the radius of orbit), the other values (such as the velocity) will compensate to preserve the value of angular momentum.

Morgan, Thomas Hunt (1866–1945)
US geneticist who helped establish that the **genes** are located on the **chromosomes**, discovered sex chromosomes, and invented the techniques of genetic mapping. He was the first to work on the fruit fly *Drosophila*, which has since become a major subject of genetic studies. He was awarded the Nobel Prize for Physiology or Medicine in 1933.

mutation
A change in the **genes** produced by a change in the **DNA** that makes up the hereditary material of all living organisms. Mutations, the raw material of **evolution**, result from mistakes during replication (copying) of DNA **molecules**. Not all mutations affect the organism, because there is a certain amount of redundancy in the genetic information. If a mutation is 'translated' from DNA into the protein that makes up the organism's structure, it may be in a nonfunctional part of the **protein** and thus have no detectable effect. This is known as a neutral mutation. Mutations affecting genes that control protein production or functional parts of protein are usually lethal to the organism.

The majority of mutations are harmful and only the few that confer any benefit on the organism are therefore favoured by **natural selection**. Mutation rates are increased by certain chemicals and by **radiation**.

N

natural selection
The process whereby the genetic make-up of a **population** changes over time as a result of certain individuals producing more descendants than others because they are better fitted to survive and reproduce in their **environment**. The accumulated effect of natural selection is to produce **adaptations** such as the insulating coat of a polar bear or the spadelike forelimbs of a mole and the emergence of new **species**. The process is slow, relying firstly on random **variation** in the **genes** of an organism being produced by **mutation** and secondly on the genetic recombination of sexual reproduction. It was recognized by Charles **Darwin** and English naturalist Alfred Russel **Wallace** as the main process driving **evolution**.

nature-nurture controversy
The long-standing dispute among philosophers and psychologists over the relative importance of **environment**, that is, upbringing, experience, and learning ('nurture'), and **heredity**, that is, genetic inheritance ('nature'), in determining the make-up of an organism, particularly as related to human personality and intelligence. One area of contention is the reason for differences between individuals in performing intelligence tests, for example. The environmentalist position assumes that individuals do not differ significantly in their inherited mental abilities and that subsequent differences are due to learning, or to differences in early experiences. Opponents insist that certain differences in the capacities of individuals (and hence their behaviour) can be attributed to inherited differences in their genetic make-up.

neo-Darwinism
The modern theory of **evolution**, built up since the 1930s by integrating Charles **Darwin's** theory of evolution through natural selection with the theory of genetic inheritance founded on the work of Gregor **Mendel**. Neo-Darwinism asserts that evolution takes place because the **environment** is slowly changing, exerting a selection pressure on the individuals within a population. Those with characteristics that happen to be fitted to the new

environment are more likely to survive and have offspring and hence pass on these favourable characteristics. Over time the genetic make-up of the **population** changes and ultimately a new **species** is formed.

neutron

One of the three main subatomic particles, the others being the **proton** and the **electron**. Neutrons have about the same **mass** as protons but no **electric charge**. They occur in the nuclei of all **atoms** except hydrogen contributing to the mass of atoms without affecting their chemistry. The **isotopes** of a single **element** differ only in the number of neutrons in their nuclei but have identical chemical properties.

The neutron is a composite particle, being made up of three **quarks**. Outside a nucleus, a free neutron is unstable, decaying with a **half-life** of 11.6 minutes into a proton, an **electron**, and an antineutrino. The neutron was discovered by the English physicist James Chadwick (1891–1974) in 1932.

Newton, Isaac (1642–1727)

English physicist and mathematician who laid the foundations of **physics** as a modern discipline. He discovered the binomial theorem, differential and integral calculus, and that white **light** is composed of many **colours**. He developed the three standard laws of motion (see **Newton's laws of motion**) and the universal law of gravitation. Newton calculated the Moon's motion on the basis of his theory of gravity and also found that his theory explained the laws of planetary motion that had been derived by German astronomer Johannes Kepler (1571–1630) on the basis of observations of the planets. His greatest achievement was to demonstrate that scientific principles are of universal application. He clearly defined the nature of mass, **weight**, force, inertia, and acceleration. His *Principia*, which includes the laws of motion, is considered one of the greatest works of science ever written.

> ❝I do not know what I may appear to the world; but to myself I seem to have been only like a boy playing on the seashore… whilst the great ocean of truth lay all undiscovered before me.❞
>
> **Isaac Newton**, quoted in L T More *Isaac Newton*

NEWTON'S LIFE

1642 Newton is born in Lincolnshire, UK on Christmas Day by the old Julian calendar (4 January 1943 by modern reckoning).
1661 He is admitted to Trinity College, Cambridge.
1669 He is elected Professor of Mathematics at Trinity, at the age of just 26.
1672 Newton is elected Fellow at the Royal Society in London.
1687 He publishes his masterpiece *Philosophae Naturalis Principia Mathematica/Mathematical Principles of Natural Philosophy*.
1689 He is elected Member of Parliament for Cambridge University.
1703 Newton is elected President of the Royal Society.
1704 He writes *Opticks*, a summary of his life's work.
1705 He is knighted by Queen Anne.
1727 Newton dies on 20 March, at the age of 80, and is buried in Westminster Abbey.

Newton's laws of motion

Three laws that form the basis of Newtonian mechanics.

1 Unless acted upon by an unbalanced force, a body at rest stays at rest, and a moving body continues moving at the same speed in the same straight line.

2 An unbalanced force applied to a body gives it an acceleration proportional to the force (and in the direction of the force) and inversely proportional to the mass of the body.

3 When a body A exerts a force on a body B, B exerts an equal and opposite force on A; that is, to every action there is an equal and opposite reaction.

> *Nature was to him an open book, whose letters he could read without effort.*
>
> **Albert Einstein** on Isaac Newton

niche

The 'place' occupied by a species in its **habitat**, including all chemical, physical, and biological components, such as what it eats, the time of day at which the **species** feeds, temperature, moisture, the parts of the habitat that it uses (for example, trees or open grassland), the way it reproduces, and how it behaves. If its habitat is considered to be 'home' to the species then its 'niche' might be considered as it 'job' there.

> **NICHE HOLDERS**
>
> It is believed that no two species can occupy exactly the same niche, because they would be in direct **competition** for the same resources at every stage of their **life cycle**.

nitrogen cycle

The path that nitrogen follows as it passes through the **ecosystem**. Nitrogen, in the form of inorganic **compounds** (such as nitrates) in the soil, is

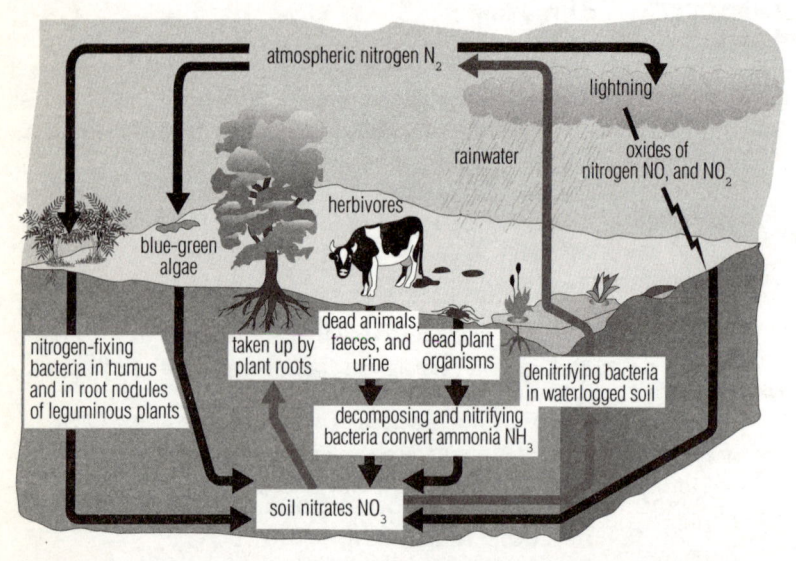

nitrogen cycle *The nitrogen cycle is one of a number of cycles during which the chemicals necessary for life are recycled. The carbon, sulphur, and phosphorus cycles are others. Since there is only a limited amount of these chemicals in the Earth and its atmosphere, the chemicals must be continuously recycled if life is to go on.*

absorbed by **plants** and turned into organic compounds (such as **proteins**) in plant tissue. A proportion of this nitrogen is incorporated into **herbivores** as they consume the plants and is in turn passed on to the **carnivores** that feed on the herbivores. The nitrogen is ultimately returned to the soil as excrement and when organisms die and are converted back to inorganic form by **decomposers**.

Although about 78% of the atmosphere is nitrogen it cannot be used directly by most organisms. Certain **bacteria** and cyanobacteria are capable of nitrogen fixation (using it to form compounds with other elements) and some of these bacteria live mutually with leguminous plants (peas and beans) and some others where they form characteristic nodules on the roots. The presence of such plants increases the nitrate content, and hence the fertility, of the soil.

nonmetal

One of a set of **elements** (around twenty in total) with certain physical and chemical properties that are generally opposite to those of **metals**. Nonmetals accept **electrons** and are sometimes called electronegative elements.

nuclear energy

Energy released from the inner core, or **nucleus**, of the **atom**. Energy produced by nuclear **fission** (the splitting of uranium or plutonium nuclei) has been harnessed since the 1950s to generate **electricity**, and research continues into the possible controlled use of nuclear fusion (the fusing, or combining, of atomic nuclei). In nuclear power stations, fission takes place in a nuclear reactor. The nuclei of uranium or, more rarely, plutonium are induced to split, releasing large amounts of heat energy. The **heat** is then removed from the core of the reactor by circulating gas or water, and used to produce the steam that drives alternators and turbines to generate electrical power.

nuclear physics

The study of the properties of the **nucleus** of the **atom**, including the structure of nuclei; nuclear forces; the interactions between particles and nuclei; and the study of radioactive decay. The study of elementary particles is **particle physics**.

nucleus (biology)

The central part of a **eukaryotic** cell that contains the cell's **DNA**. The **nucleus** controls the function of the **cell** by determining which **proteins** are produced within it. The nucleus contains the nucleolus, the part of the cell where ribosomes (responsible for protein assembly) are produced. Movement of molecules into and out of the nucleus occurs through pores in the nuclear membrane. An average mammalian nucleus has approximately 3,000 pores.

nucleus (physics)

The positively charged central part of an **atom**, which constitutes almost all its **mass**. Except for hydrogen nuclei, which have only **protons**, nuclei are composed of both protons and **neutrons**. Surrounding the nuclei are **electrons**, of equal and opposite charge to that of the protons, thus giving the atom a neutral charge.

RUTHERFORD'S DISCOVERY

The nucleus was discovered by New Zealand-born British physicist Ernest Rutherford (1871–1937) in 1911. Firing alpha particles at a sheet of gold foil, Rutherford noticed that a few of the particles were deflected back. Astonished, he remarked: 'It was almost as if you fired a 15-inch shell at a piece of tissue paper and it came back and hit you!' The deflection, he deduced, was due to the positively charged alpha particles being repelled by approaching a small but dense positively charged nucleus.

Oersted, Hans Christian (1777–1851)
Danish physicist who founded the science of **electromagnetism**. In 1820 he discovered the magnetic field associated with an **electric current**. He had predicted in 1813 that an electric current would produce **magnetism** when it flowed through a wire, just as it produced **heat** and **light**. His 1820 experiment involved a compass needle placed beneath a wire connected to a **battery**. He found that a circular magnetic field is produced around a wire carrying a current. This paved the way for Michael **Faraday's** invention of the electric motor.

omnivore
An animal that feeds on both **plant** and **animal** material. Omnivores have digestive adaptations intermediate between those of **herbivores** and **carnivores**, with relatively unspecialized digestive systems and gut **microorganisms** that can digest a variety of foodstuffs. Omnivores include humans, the chimpanzee, the cockroach, and the ant.

oncogene
A **gene** carried by a **virus** that induces a cell to divide abnormally, giving rise to a **cancer**. Oncogenes arise from **mutations** in genes (proto-oncogenes) found in all normal **cells**. They are usually also found in **viruses** that are capable of transforming normal cells to tumour cells. Such viruses are able to insert their oncogenes into the host cell's **DNA**, causing it to divide uncontrollably. More than one oncogene may be necessary to transform a cell in this way.

Oppenheimer, J(ulius) Robert (1904–1967)
US physicist who, as director of the Los Alamos Science Laboratory 1943–1945 was in charge of the development of the Manhattan Project to develop the atom bomb. When later he realized the dangers of radioactivity, he objected to the development of the hydrogen bomb, and was alleged to be a security risk in 1953 by the US Atomic Energy Commission (AEC). Investigating the equations describing the energy states of the atom,

Oppenheimer showed in 1930 that a positively charged particle with the mass of an electron could exist. This particle was detected in 1932 and called the positron.

optics
The branch of physics that deals with the study of light and vision – for example, shadows and mirror images, lenses, microscopes, telescopes, and cameras.

osmosis
The movement of **water** through a selectively permeable membrane separating **solutions** of different concentrations. Water passes by **diffusion** from a weak solution (high water concentration) to a strong solution (low water concentration) until the two concentrations are equal. The selectively permeable membrane allows the **diffusion** of water but not of the solute (for example, sugar molecules). Many cell membranes behave in this way, and osmosis is a vital mechanism in the transport of fluids in living organisms – for example, in the transport of water from soil (weak solution) into the roots of plants (stronger solution of cell sap).

Osmoregualtion
Excessive flow of water into a cell by osmosis can damage it. Cells protect against this using processes of osmoregulation. The cell wall of a fully turgid cell exerts **pressure** on the solution within the cell and osmosis is arrested. By this mechanism plant cells can osmoregulate. **Fish** have protective mechanisms in the kidney to counteract osmosis, which would otherwise cause fluid transport between the body of the animal and the surrounding water (outwards in saltwater fish, inwards in freshwater ones).

oxidation
The loss of **electrons**, gain of **oxygen**, or loss of hydrogen by an **atom**, **ion**, or **molecule** during a chemical **reaction**. Oxidation may be brought about by reaction with another **compound** (oxidizing agent), which simultaneously undergoes **reduction**, or electrically at the **anode** (positive electrode) of an electrolytic cell.

oxygen
A colourless, odourless, tasteless, nonmetallic, gaseous **element**. Oxygen is very reactive and combines with all other elements except the **inert gases** and

fluorine. Oxygen exists in molecular form, either O_2 or O_3 (**ozone**), rather than as free atoms. It is present in carbon dioxide, silicon dioxide (quartz), iron ore, and calcium carbonate (limestone). Oxygen is a by-product of **photosynthesis** and the basis for **respiration** in plants and animals.

Oxygen is the most abundant element in the Earth's crust (almost 50% by mass), forms about 21% by volume of the atmosphere, and is present in combined form in water and many other substances.

ozone

A highly reactive pale-blue **gas** with a penetrating odour. Ozone is an oxygen **molecule** that is made up of three **atoms** of **oxygen** (O_3) rather than the more usual two. It is formed when O_2 oxygen molecules are split by ultraviolet radiation or electrical discharge. It forms the ozone layer in the upper atmosphere, which protects life on Earth from the Sun's **ultraviolet** rays, a cause of skin **cancer**.

P

parasite
An organism that lives on or in another organism (called the host) and depends on it for nutrition, often at the expense of the host's welfare. Parasites that live inside the host, such as liver flukes and tapeworms, are called endoparasites; those that live on the exterior, such as fleas and lice, are called ectoparasites. Parasitic wasps, such as ichneumons, are more correctly termed parisitoids, as they ultimately kill their hosts.

parthenogenesis
The development of an ovum (egg) without any genetic contribution from a male. Parthenogenesis is the normal means of reproduction in a few plants (for example, dandelions) and animals (for example, certain fish). Some sexually reproducing species, such as aphids, show parthenogenesis at some stage in their **life cycle** to accelerate reproduction to take advantage of good conditions.

particle physics
The study of the particles that make up all **atoms**, and of their interactions. More than 300 subatomic particles have now been identified by physicists, categorized into several classes according to a veriety of properites such as their **mass**, **electric charge**, spin, and interactions.

Subatomic particles include the elementary particles, such as **quarks**, which are believed to be indivisible and so may be considered the fundamental units of **matter**; and particles such as the proton and neutron, which are composite particles, made up of two or three quarks. The **proton**, **electron**, and neutrino are the only stable particles (the **neutron** being stable only when in the atomic **nucleus**). The unstable particles, discovered through experiments with particle accelerators and cosmic radiation, decay rapidly into other particles.

Pasteur, Louis (1822–1895)
French chemist and microbiologist who discovered that **fermentation** is caused by **micro-organisms** and developed the germ theory of **disease**. He

also created a vaccine for rabies, which led to the foundation of the Pasteur Institute in Paris in 1888.

Fermentation and pasteurization
Pasteur's research into fermentation was prompted by a query from an industrialist about wine and beer-making. He proved that fermentation does not require **oxygen**, yet it involves living micro-organisms, and that, to produce the desired type of fermentation it is necessary to use the correct type of yeast. Pasteur also discovered that if wine is heated to about 50°C/122°F this kills the yeast and thereby prevents souring during the ageing process. This sterilizing process is now called pasteurization.

Pasteur *French chemist and microbiologist Louis Pasteur, whose germ theory of disease led to radical improvements in medical hygiene and provided the rationale for immunization.*

Spontaneous generation
For centuries it had been believed that organisms could appear spontaneously from decaying matter. Pasteur showed that dust in the air contains spores of living organisms that reproduce when they come into contact with a suitable food source. If the food was protected from contamination it did not go bad.

The germ theory
Pasteur combined his discoveries to form the germ theory of disease, possibly the single most important single medical discovery of all time, because it provided both a practical method of combating disease by disinfection and a theoretical foundation for further research.

PASTEUR'S LIFE

1822 Pasteur is born on 27 December in Dole in eastern France.
1843 He enters the Ecole Normale Superieure where he begins to study chemistry.

1848	He is appointed Professor of Physics at Dijon Lycee.
1849	Pasteur accepts the post of Professor of Chemistry at the University of Strasbourg.
1863	He is elected to a chair created for him at the Ecole Normale Superieure and institutes an original teaching programme relating chemistry, physics, and geology to the fine arts. He also becomes Dean of the new science faculty at Lille University where he institutes the novel concept of evening classes for workers.
1868	Pasteur identifies a minute parasite affecting the French silk industry, he recommends that all the infected silk worms should be destroyed and the disease is eliminated.
1881	He develops a method for reducing the virulence of certain micro-organisms, by heating.
1885	Pasteur successfully inoculates a young boy with his rabies vaccine.
1888	He heads the Pasteur Institute in Paris, which is created for the purpose of continuing his research into rabies.
1895	Pasteur dies on 28 September, in Paris.

pathogen

Any **micro-organism** that causes **disease**. Most pathogens are **parasites**, and the diseases they cause are incidental to their search for food or shelter inside the host. Nonparasitic organisms, such as soil **bacteria** or those living in the human gut and feeding on waste foodstuffs, can also become pathogenic to a person whose immune system or liver is damaged.

periodic table of the elements

A table in which the **elements** are arranged in order of their atomic number (the number of **protons** in the **nucleus**). The table summarizes the major properties of the elements and enables predictions to be made about their behaviour. These features are a direct consequence of the electronic and nuclear structure of the **atoms** of the elements. The relationships established between the positions of elements in the periodic table and their major properties has enabled scientists to predict the properties of previously undiscovered elements – for example, technetium, atomic number 43, first synthesized in 1937.

Groups

The elements in the periodic table are divided into groups (vertical columns) numbered I–VII and 0 to reflect the number of **electrons** in the atom's outermost shell, and hence its **valency**. There are striking similarities in the chemical properties of the elements in each of the groups. Reactivity increases down the group. A large block between groups II and III contains the transition elements, characterized by displaying more than one valency state.

Periods

A gradation of properties may be traced along the horizontal rows (called periods) of the table. Metallic character increases across a period from right to left, and down a group. (*See table on pp 152–153*.)

perpetual motion

The idea that a machine can be designed and constructed in such a way that, once started, it will continue in motion indefinitely without requiring any further input of **energy** (motive power). Such a device would contradict at least one of the laws of **thermodynamics**. As a result, all practical machines require a continuous supply of energy, and no heat engine is able to convert all the **heat** it produces into useful work.

pH

A scale for measuring acidity or alkalinity. pH is a measure of the number of hydrogen **ions** that enter **solution** when a substance is dissolved in **water**. The lower the pH number the greater the number of hydrogen ions present. A pH of 7.0 indicates neutrality, below 7 is acidic, while above 7 is alkaline. Strong **acids**, such as those used in car batteries, have a pH of about 2; strong **alkalis** such as sodium hydroxide are pH 13.

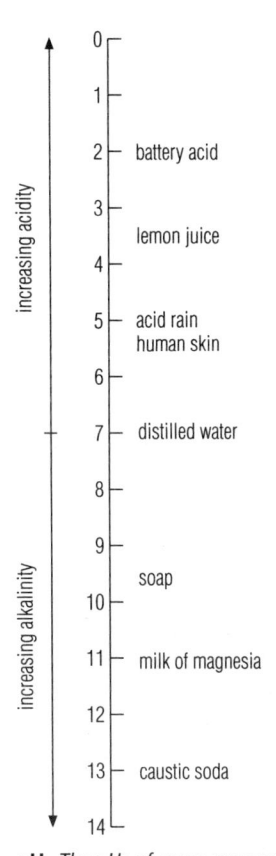

```
increasing acidity
 0
 1
 2  battery acid
 3
 4  lemon juice
 5  acid rain
    human skin
 6
 7  distilled water
 8
increasing alkalinity
 9
10  soap
11  milk of magnesia
12
13  caustic soda
14
```

pH *The pHs of some common substances. The lower the pH is, the more acidic the substance; the higher the pH, the more alkaline the substance.*

periodic table of the elements

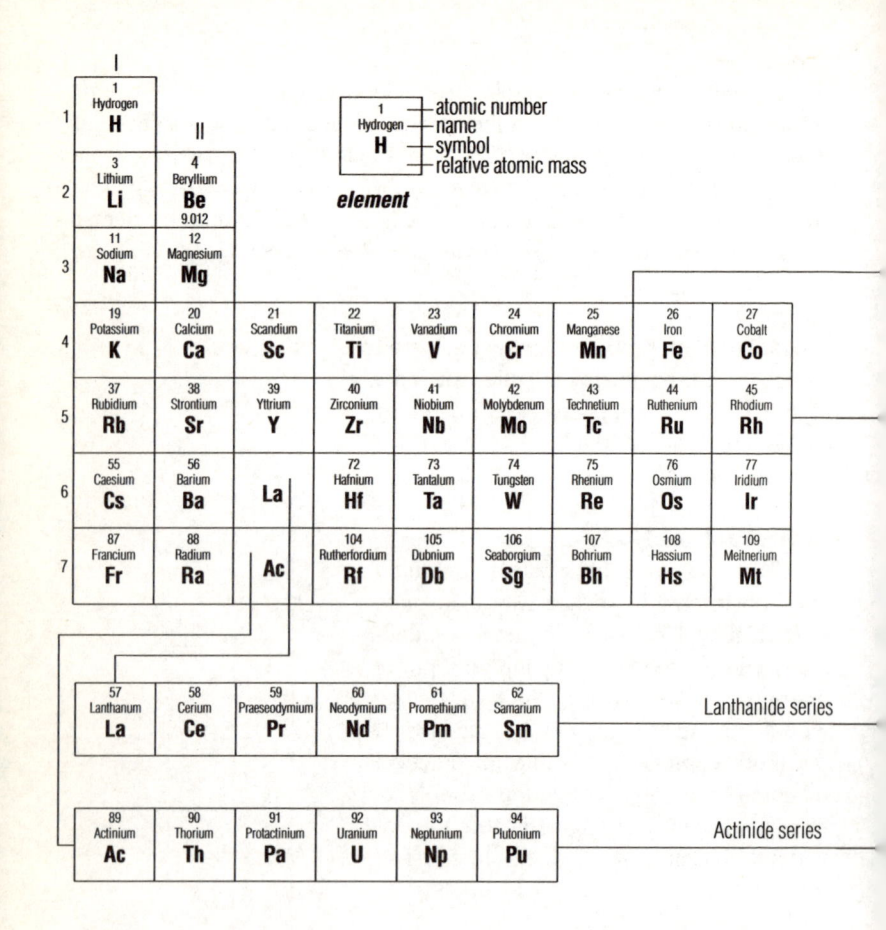

	I									
1	1 Hydrogen **H**	II								
2	3 Lithium **Li**	4 Beryllium **Be** 9.012								
3	11 Sodium **Na**	12 Magnesium **Mg**								
4	19 Potassium **K**	20 Calcium **Ca**	21 Scandium **Sc**	22 Titanium **Ti**	23 Vanadium **V**	24 Chromium **Cr**	25 Manganese **Mn**	26 Iron **Fe**	27 Cobalt **Co**	
5	37 Rubidium **Rb**	38 Strontium **Sr**	39 Yttrium **Y**	40 Zirconium **Zr**	41 Niobium **Nb**	42 Molybdenum **Mo**	43 Technetium **Tc**	44 Ruthenium **Ru**	45 Rhodium **Rh**	
6	55 Caesium **Cs**	56 Barium **Ba**	**La**	72 Hafnium **Hf**	73 Tantalum **Ta**	74 Tungsten **W**	75 Rhenium **Re**	76 Osmium **Os**	77 Iridium **Ir**	
7	87 Francium **Fr**	88 Radium **Ra**	**Ac**	104 Rutherfordium **Rf**	105 Dubnium **Db**	106 Seaborgium **Sg**	107 Bohrium **Bh**	108 Hassium **Hs**	109 Meitnerium **Mt**	

Element key: atomic number, name, symbol, relative atomic mass.

Lanthanide series:
| 57 Lanthanum **La** | 58 Cerium **Ce** | 59 Praesodymium **Pr** | 60 Neodymium **Nd** | 61 Promethium **Pm** | 62 Samarium **Sm** |

Actinide series:
| 89 Actinium **Ac** | 90 Thorium **Th** | 91 Protactinium **Pa** | 92 Uranium **U** | 93 Neptunium **Np** | 94 Plutonium **Pu** |

periodic table of the elements *The periodic table of the elements arranges the elements into horizontal rows (called periods) and vertical columns (called groups) according to their atomic numbers. The elements in a group or column all have similar properties – for example, all the elements in the far right-hand column are inert gases.*

pheromone

A chemical signal (such as an odour) that is emitted by one animal and affects the behaviour of others. Pheromones are used by many animal species to attract mates.

				III	IV	V	VI	VII	0
									2 Helium **He**
				5 Boron **B**	6 Carbon **C**	7 Nitrogen **N**	8 Oxygen **O**	9 Fluorine **F**	10 Neon **Ne**
				13 Aluminium **Al**	14 Silicon **Si**	15 Phosphorus **P**	16 Sulphur **S**	17 Chlorine **Cl**	18 Argon **Ar**
28 Nickel **Ni**	29 Copper **Cu**	30 Zinc **Zn**	31 Gallium **Ga**	32 Germanium **Ge**	33 Arsenic **As**	34 Selenium **Se**	35 Bromine **Br**	36 Krypton **Kr**
46 Palladium **Pd**	47 Silver **Ag**	48 Cadmium **Cd**	49 Indium **In**	50 Tin **Sn**	51 Antimony **Sb**	52 Tellurium **Te**	53 Iodine **I**	54 Xenon **Xe**
78 Platinum **Pt**	79 Gold **Au**	80 Mercury **Hg**	81 Thallium **Tl**	82 Lead **Pb**	83 Bismuth **Bi**	84 Polonium **Po**	85 Astatine **At**	86 Radon **Rn**
110 Unennilium **Uun**	111 Unununium **Uuu**	112 Ununbium **Uub**

63 Europium **Eu**	64 Gadolinium **Gd**	65 Terbium **Tb**	66 Dysprosium **Dy**	67 Holmium **Ho**	68 Erbium **Er**	69 Thulium **Tm**	70 Ytterbium **Yb**	71 Lutetium **Lu**

95 Americium **Am**	96 Curium **Cm**	97 Berkelium **Bk**	98 Californium **Cf**	99 Einsteinium **Es**	100 Fermium **Fm**	101 Mendelevium **Md**	102 Nobelium **No**	103 Lawrencium **Lr**

photon

An elementary particle consisting of a quantum (the smallest possible amount) of electromagnetic energy, such as light or **X-rays**. The photon has both particle and wave properties; it has no charge, and is considered massless but possesses **momentum** and **energy**. It is the carrier of the electromagnetic force, one of the **fundamental forces** of nature.

See also: *Max Planck, quantum theory.*

photosynthesis

The process by which green plants trap **light** energy from the Sun. This energy is used to drive a series of chemical **reactions** which lead to the formation of glucose, which provides the basic food for both plants and animals. For photosynthesis to occur, the plant must possess **chlorophyll** and must have a supply of carbon dioxide and water. Photosynthesis takes place inside chloroplasts, which are found mainly in the leaf cells of **plants**. The chemical reactions of photosynthesis occur in two stages.

- *The light reaction:* sunlight is used to split water (H_2O) into **oxygen** (O_2), **protons** (hydrogen **ions**, H^+), and **electrons**.
- *The dark reaction*: the protons and electrons are used to convert carbon dioxide (CO_2) into **carbohydrates**.

PHOTOSYNTHESIS AND OXYGEN

The by-product of photosynthesis, oxygen, is of great importance to almost all living organisms, and virtually all atmospheric oxygen has originated by photosynthesis.

---

physics
The branch of science concerned with the laws that govern the structure of the universe, and the investigation of the properties of **matter** and **energy** and their interactions. For convenience, physics is often divided into branches such as atomic physics, nuclear physics, **particle physics**, solid-state physics, molecular physics, **electricity** and **magnetism**, **optics**, **acoustics**, **heat**, **thermodynamics**, **quantum theory**, and **relativity**.

Planck, Max Karl Ernst (1858–1947)
German physicist who framed the **quantum theory** in 1900. His research into the manner in which heated bodies radiate energy led him to report that **energy** is emitted only in discrete amounts, called 'quanta' (singular quantum), the magnitudes of which are proportional to the frequency of the radiation. His discovery ran counter to classical physics and is held to have marked the commencement of the modern science. He was awarded the Nobel Prize for Physics in 1918.

plankton
Small, often microscopic, forms of **plant** and **animal** life that live in the upper layers of fresh and salt water, and are an important source of **food** for

larger animals. Marine plankton is concentrated in areas where rising currents bring mineral salts to the surface.

plant

An organism that makes its own food from inorganic materials through the process of **photosynthesis**. Plants are the primary producers in almost all **food chains**, so that all animal life is ultimately dependent on them. They play a vital part in the **carbon cycle**, removing carbon dioxide from the atmosphere and generating **oxygen**. The study of plants is known as **botany**.

> **PREDATORY PLANKTON**
>
> *Pfiesteria piscicida* stuns its prey by producing a powerful toxin that also causes the deaths of nearby fish and may be harmful to humans exposed to it.

Plant properties
- photosynthesis
- rigid cellulose cell walls
- immobility.

The most advanced plants are the vascular plants, which have special supportive water-conducting tissues. This group includes all **seed** plants, that is the **gymnosperms** (conifers, yews, cycads, and ginkgos) and the **angiosperms** (flowering plants). The seed plants are the largest plant group. Vascular plants are usually divided into three parts: root, stem, and leaves. Stems grow above or below ground. They carry **water** and **salts** from the roots to the leaves and sugars from the leaves to the roots. The leaves manufacture the food of the plant by means of photosynthesis. Flowers and cones are modified leaves arranged in groups, enclosing the reproductive organs from which the fruits and seeds result.

> **PLANT ANCESTOR**
>
> *Cooksonia pertoni*, a tiny **fossil** plant only a few centimetres high and 400 million years old, is considered to be the ancestor of all vascular plants.

plasma (physics)

An ionized **gas** produced at extremely high **temperatures**, such as those found in the Sun and other stars, which contains positive and negative charges in equal numbers. It is a good electrical conductor. The plasma produced in thermonuclear reactions is confined through the use of magnetic fields.

plasma (biology)
The liquid component (90% water) of blood in which a number of substances are dissolved. These include a variety of proteins (around 7%), inorganic mineral salts such as sodium and calcium, waste products such as urea, traces of hormones, and antibodies to defend against infection.

pollination
The process by which pollen is transferred from one **plant** to another. The male sex cells are contained in pollen grains, which must be transferred from the anther to the stigma in **angiosperms** (flowering plants), and from the male cone to the female cone in **gymnosperms** (cone-bearing plants). **Fertilization** (which is not the same as pollination) occurs after the growth of the pollen tube to the ovary. Self-pollination occurs when pollen is transferred to a stigma of the same flower, or to another flower on the same plant; cross-pollination occurs when pollen is transferred to another plant. This involves external pollen-carrying agents, such as wind, water, insects, birds, bats, and other small mammals.

> **MUTUAL BENEFIT**
>
> Plants that rely on animals to pollinate them generally produce nectar, a sugary liquid, or surplus pollen, or both, on which the pollinator feeds. Thus the relationship between pollinator and plant is an example of mutualism, in which both benefit.

polymer
A compound made up of a large long-chain or branching matrix composed of many repeated simple molecules linked together by a chemical process called polymerization. There are many polymers, both natural (cellulose) and synthetic (polyethylene and nylon, types of plastic).

polyunsaturate
A type of **fat** or oil, such as vegetable and fish oils, which are **liquids** at room temperature. Polyunsaturates contain a high proportion of triglyceride

molecules whose fatty acid chains contain several double **bond**s linking the **carbon** atoms in the molecule. The more double bonds the fatty-acid chains contain, the lower the **melting point** of the fat. By contrast, saturated fats, with no double bonds, are **solids** at room temperature. The polyunsaturated fats used for margarines are produced by taking a vegetable or fish oil and turning some of the double bonds to single bonds, so that the product is semi-solid at room temperature. Medical evidence suggests that polyunsaturated fats, used widely in margarines and cooking fats, are less likely to contribute to cardiovascular disease than saturated fats, but there is also some evidence that they may have adverse effects on health.

population
A group of animals or plants of one **species**, living in a certain area and able to interbreed; the members of a given species in a **community** of living things.

potential energy
Energy possessed by an object by virtue of its relative position or state (for example, as in a compressed spring or a muscle). It is contrasted with **kinetic energy**, the form of energy possessed by moving bodies. An object that has been raised up is described as having gravitational potential energy.

power
The rate of doing work or consuming **energy**. It is measured in watts (joules per second) or other units of work per unit time.

pressure
The force acting normally on a surface or the ratio of force to area. The **SI** unit of pressure is the pascal (Pa), equal to a pressure of one newton per square metre. In the atmosphere, the pressure declines with height from about 100 kPa at sea level to zero where the atmosphere fades into space. Pressure is commonly measured with a barometer, manometer, or Bourdon gauge.

primate
Any member of the order of **mammals** that includes monkeys, apes, and humans (together called anthropoids), as well as lemurs, bushbabies, lorises, and tarsiers

> **PRIMATE POPULATION**
>
> Brazil has 77 species of primate, far more than any other country.

(together called prosimians). Generally, they have forward-directed eyes, gripping hands and feet, opposable thumbs, and big toes; they tend to have nails rather than claws, with gripping pads on the ends of the digits, all adaptations to the arboreal, climbing mode of life.

prion

A prion is an infectious agent a hundred times smaller than a **virus**. Composed of a simple **protein**, and without any detectable genetic material, prions are strongly linked to a number of fatal degenerative brain **diseases** in **mammals**, such as bovine spongiform encephalopathy (BSE) in cattle, scrapie in sheep, and Creutzfeldt–Jakob disease (CJD) and kuru in humans.

The gene for the prion protein is found normally in all mammals, including humans. The protein can exist in two forms: a normal, harmless form (found in white blood cells and on the surface of brain nerve cells) and an abnormal, disease-causing form that causes a **chain reaction** that induces the normal protein to fold into the abnormal form. As the prions accumulate in the nervous system the tissues are damaged, creating holes in the brain and giving it a characteristic spongy appearance.

> **PRUSINER'S PRIONS**
>
> The existence of prions was postulated by US neurologist Stanley Prusiner (1942–) in 1982, when he and his colleagues isolated a single infectious agent for scrapie that consisted only of protein and had no associated nucleic acid.

prokaryote

An organism whose **cells** lack organelles (specialized segregated structures such as nuclei, mitochondria, and chloroplasts). Prokaryote **DNA** is not arranged in **chromosomes** but forms a coiled structure called a nucleoid. The prokaryotes include the **bacteria, archaea**, and cyanobacteria; all other organisms are **eukaryotes**.

protein

One of a large group of complex, biologically important, organic **compounds** found in all living organisms. Proteins are composed of long chains of **amino acids** folded into characteristic shapes. As **enzymes** proteins regulate all aspects of **metabolism**. Structural proteins such as keratin and

amino acids, where R is one of many possible side chains

peptide – this is one made of just three amino acid units.

protein *A protein molecule is a long chain of amino acids linked by peptide bonds. The properties of a protein are determined by the order, or sequence, of amino acids in its molecule, and by the three-dimensional structure of the molecular chain. The chain folds and twists, often forming a spiral shape.*

collagen make up the skin, claws, bones, tendons, and ligaments; muscle proteins produce movement; haemoglobin transports **oxygen**; and membrane proteins regulate the movement of substances into and out of **cells**.

protist

A single-celled organism which has a **eukaryotic** cell, but which is not a member of the **plant**, **fungal**, or **animal** kingdoms. The main protists are protozoa. The term is often used for members of the kingdom Protista, which features in certain **classifications** of the living world. This kingdom may include slime moulds, all **algae** (seaweeds as well as unicellular forms), and protozoa.

proton

A positively charged subatomic particle, a constituent of the **nucleus** of all **atoms**. A proton is extremely long-lived, with a lifespan of at least 10^{32} years. It carries a unit positive charge equal to the negative charge of an **electron**. Its mass is almost 1,836 times greater than that of an electron. The number of protons in the atom of an element is equal to the atomic number of that **element**.

Q

quantum theory or quantum mechanics
The theory that **energy** does not have a continuous range of values, but is, instead, absorbed or radiated discontinuously, in multiples of definite, indivisible units called quanta. Just as earlier theory showed how light, generally seen as a wave motion, could also in some ways be seen as composed of discrete particles (photons), quantum theory shows how atomic particles such as electrons may also be seen as having wavelike properties. Quantum theory is the basis of particle physics, modern theoretical chemistry, and the solid-state physics that describes the behaviour of the silicon chips used in computers.

QUANTUM PIONEERS

The theory began with the work of Max **Planck** in 1900 on radiated energy, and was extended by Albert **Einstein** to electromagnetic radiation generally, including light. Later work by Erwin **Schrodinger**, Werner **Heisenberg**, Paul Dirac, and others elaborated the theory to what is called quantum mechanics (or wave mechanics).

quark

The elementary particle that is the fundamental constituent of subatomic particles such as **neutrons** and **protons**. Quarks have **electric charges** that are fractions of the electronic charge ($+\frac{2}{3}$ or $-\frac{1}{3}$ of the electronic charge). There are six types, or 'flavours': up, down, top, bottom, strange, and charmed, each of which has three varieties, or 'colours': red, green, and blue (visual colour is not meant, although the analogy is useful in many ways). For each quark there is an antiparticle, called an antiquark.

TOP QUARK

The existence of the top quark was confirmed by two teams of physicists at Fermilab in March 1995. It is unstable and unlikely to last more than a millionth of a billionth of a billionth of a second, but experiments at Fermilab suggest it is at least as massive as a silver atom.

radiation

Energy travelling in the form of **photons** or electromagnetic **waves**, for example **heat** or **light**; or a stream of particles from a radioactive source, particularly alpha particles and beta particles.

radioactivity

The spontaneous disintegration of the nuclei of radioactive atoms, accompanied by the emission of **radiation** in the form of alpha particles, beta particles or gamma rays (see **gamma radiation**). It is a property of all radioactive **isotopes**, and can be either natural or induced.

- *Alpha particles:* positively charged, high-energy particles emitted from the **nucleus** of a radioactive **atom**. They consist of two **neutrons** and two **protons** and are thus identical to the nucleus of a helium atom. Because of their large mass, alpha particles have a short range of only a few centimetres in air, and can be stopped by a sheet of paper.

- *Beta particles*: high-energy, high velocity **electrons**, more penetrating than alpha rays and can travel through a 3-mm/0.1-in sheet of aluminium or up to 4 cm/1.5 in of air.

- *Gamma rays*: very high-**frequency** electromagnetic radiation stopped only by direct collision with an atom and therefore very penetrating; can be stopped by about 1 m/3.3 ft of lead.

Ionization

When alpha, beta, and gamma radiation pass through matter they tend to knock electrons out of atoms, ionizing them. They are therefore called ionizing radiation. Alpha particles are the most ionizing, being heavy, slow moving and carrying two positive charges. Gamma rays are weakly ionizing as they carry no charge. Beta particles fall between alpha and gamma radiation in their ionizing potential.

radio astronomy

The study of radio waves emitted naturally by objects in space, by means of a radio telescope. Radio emission comes from hot gases (thermal radiation); **electrons** spiralling in magnetic fields (synchrotron radiation); and specific wavelengths (lines) emitted by atoms and **molecules** in space. Radio astronomy began in 1932 when US astronomer Karl Jansky detected radio waves from the centre of our galaxy, but the subject did not develop until after World War II. Radio astronomy has greatly improved our understanding of the evolution of stars, the structure of galaxies, and the origin of the **universe**.

radiocarbon dating or carbon dating

A method of dating organic materials (for example, bone or wood), used in archaeology. Plants take up carbon dioxide gas from the atmosphere and incorporate it into their tissues, and some of that carbon dioxide contains the radioactive **isotope** of **carbon,** ^{14}C or carbon-14. As this decays at a known rate (half of it decays every 5,730 years), the time elapsed since the plant died can be measured in a laboratory. Animals take carbon-14 into their bodies from eating plant tissues and their remains can be similarly dated. After 120,000 years so little carbon-14 is left that no measure is possible.

radio wave

An electromagnetic **wave** possessing a long wavelength (ranging from about 1mm to 104 m) and a low **frequency** (from about 105 to 1011 Hz). Radio waves that are used for communications have all been modulated to carry information. Certain astronomical objects emit radio waves, which may be detected and studied using radio telescopes.

Radio spectrum

- microwaves, used for both communications and for cooking
- ultra high- and very high-frequency (UHF and VHF) waves, used for television and FM (frequency modulation) radio communications
- short, medium, and long waves, used for AM (amplitude modulation) radio communications.

reaction, chemical

The coming together of two or more **atoms**, **ions**, or **molecules** with the result that a chemical change takes place. A change that occurs when two or more substances interact with each other, resulting in the production of different substances with different chemical compositions.

- *Endothermic* reactions take in **heat** from the surroundings.
- *Exothermic* reactions give out heat.
- *A reversible reaction* proceeds in both directions at the same time, as the product decomposes back into reactants as it is being produced. Such reactions do not run to completion, provided that no substance leaves the system.

reduction, chemical

The gain of **electrons**, loss of **oxygen**, or gain of hydrogen by an **atom**, **ion**, or **molecule** during a chemical **reaction**. Reduction may be brought about by reaction with another **compound**, which is simultaneously oxidized (reducing agent), or electrically at the **cathode** (negative **electrode**) of an electric **cell**. **Photosynthesis** is essentially a mechanism for reducing carbon dioxide to **carbohydrate**.

reflection

The bouncing back or deflection of **waves**, such as **light** or **sound** waves, when they hit a surface.

> **LAW OF REFLECTION**
>
> The angle of incidence (the angle between the ray and a perpendicular line drawn to the surface) is equal to the angle of reflection (the angle between the reflected ray and a perpendicular to the surface).

Total internal reflection
When light passes from a denser medium to a less dense medium, such as from water to air, both **refraction** and reflection can occur. If the angle of incidence is small, the reflection will be relatively weak compared to the refraction. But as the angle of incidence increases the relative degree of reflection will increase. At some critical angle of incidence the angle of refraction is 90°. Since refraction cannot occur above 90°, the light is totally reflected at angles above this critical angle of incidence. This condition is known as total internal reflection. Total internal reflection is used in fibre optics to transmit data over long distances, without the need of amplification.

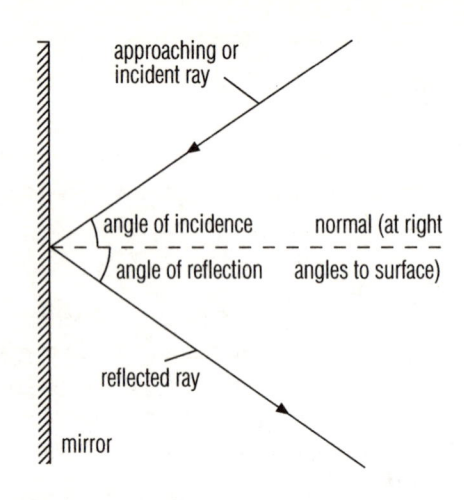

reflection *A ray of light is reflected off a mirror.*

reflex

A very rapid involuntary response to a particular stimulus. It is controlled by the nervous system. A reflex involves only a few nerve cells, unlike the slower but more complex responses produced by the many processing nerve cells of the brain. A simple reflex is entirely automatic and involves no learning. Examples of such reflexes include the sudden withdrawal of a hand in response to a painful stimulus, or the jerking of a leg when the kneecap is tapped. Sensory cells (receptors) in the knee send signals to the spinal cord along a sensory nerve cell. Within the spine a reflex arc switches the signals straight back to the muscles of the leg (effectors) via an intermediate nerve cell and then a motor nerve cell; contraction of the leg occurs, and the leg kicks upwards. Only three nerve cells are involved, and the brain is only aware of the response after it has taken place.

Conditioned reflexes

A conditioned reflex involves the modification of a reflex action in response to experience (learning). A stimulus that produces a simple reflex response becomes linked with another, possibly unrelated, stimulus. For example, a dog may salivate (a reflex action) when it sees its owner remove a tin-opener from a drawer because it has learned to associate that stimulus with the stimulus of being fed.

refraction

The bending of a wave when it passes from one medium into another. It is the effect of the different speeds of **wave** propagation in two substances that have different densities. The degree of refraction depends on the relative densities of the media, the angle at which the wave strikes the surface of the second medium, and the amount of bending and change of **velocity** corresponding to the wave's **frequency** (dispersion). Refraction occurs with all types of progressive waves – **electromagnetic** waves, **sound** waves, and water waves. It differs from **reflection**, which involves no change in velocity.

refraction *Refraction is the bending of a light beam when it passes from one transparent medium to another. This is why a spoon appears bent when standing in a glass of water and pools of water appear shallower than they really are.*

The angle between the incoming light ray and a line perpendicular to the surface (the normal) is called the angle of incidence. The angle between the refracted ray and the normal is called the angle of refraction. The angle of refraction cannot exceed 90°.

relativity

Theories concerning motion formulated by Albert **Einstein** in 1905 and 1915.

Special theory of relativity (1905)
Starting with the premises that:

- the laws of nature are the same for all observers in unaccelerated motion, and
- the **speed of light** is independent of the motion of its source

Einstein arrived at some rather unexpected conclusions. Intuitively familiar concepts, like mass, length, and time, had to be modified to accomodate the unvarying speed of light. For example, an object moving rapidly past the observer will appear to be both shorter and heavier than when it is at rest relative to the observer, and a clock moving rapidly past the observer will appear to be running slower than when it is at rest. These predictions of relativity theory are quite negligible at speeds less than about 1,500 km/s, and they only become appreciable at speeds approaching that of light.

General theory of relativity (1915)

The general theory dealt with objects in accelerated motion. This led Einstein to a consideration of **gravity**. **Space-time** was modified, or deformed, by the presence of a body with **mass**. A planet's orbit around the Sun (as observed in three-dimensional space) arises from its natural trajectory in modified space-time; there is no need to invoke, as Isaac **Newton** did, a force of gravity coming from the Sun and acting on the planet. General relativity is central to modern astrophysics and **cosmology**; predicting, for example, the possibility of **black holes**.

reptile

Any member of the vertebrates class Reptilia. Unlike **amphibians**, reptiles have hard-shelled, yolk-filled eggs that are laid on land and from which fully formed young are born. Some snakes and lizards retain their eggs and give birth to live young. Reptiles are

> **REPTILES REVEALED**
>
> A four-year study of rainforest in eastern Madagascar revealed 26 new reptile species 1995.

cold-blooded, and their skin is usually covered with scales. The **metabolism** is slow, and in some cases (certain large snakes) intervals between meals may be months.

Reptilia:
- Chelonia (tortoises and turtles)
- Crocodilia (alligators and crocodiles)
- Squamata
    Lacertilia (lizards)
    Ophidia or Serpentes (snakes)
    Amphisbaenia (worm lizards)
- Rhynchocephalia, one surviving species, the lizardlike tuatara of New Zealand.

resistance

That property of a **conductor** that restricts the flow of **electricity** through it, associated with the conversion of electrical **energy** to **heat**; also the magnitude of this property. Resistance depends on many factors, such as the nature of the material, its **temperature**, dimensions, and thermal properties; degree of impurity and the frequency and magnitude of the current. The **SI** unit of resistance is the ohm.

> **OHM'S LAW**
>
> $V = IR$
>
> V = potential difference in volts,
> I = the current in amperes
> R = resistance in ohms

resonance

The rapid amplification of a **vibration** when the vibrating object is subject to a force varying at its natural **frequency**. In a trombone, for example, the length of the air column in the instrument is adjusted until it resonates with the note being sounded. Resonance effects are also produced by many electrical circuits. Tuning a radio, for example, is done by adjusting the natural frequency of the receiver circuit until it coincides with the frequency of the radio waves falling on the aerial.

> **DESTRUCTIVE RESONANCE**
>
> Resonance caused the collapse of the Tacoma Narrows Bridge, USA, 1940, when the frequency of the wind gusts coincided with the natural frequency of the bridge.

respiration

A metabolic process in organisms in which food **molecules** are broken down to release **energy**. The **cells** of all living organisms need a continuous supply of energy, and in most plants and animals this is obtained by **aerobic** respiration. In this process, **oxygen** is used to break down the glucose molecules in food. This releases energy in the form of energy-carrying molecules (ATP), and produces carbon dioxide and water as by-products. Respiration sometimes occurs without oxygen. This is called **anaerobic** respiration. In this case, the end products are energy and either lactose acid or ethanol (**alcohol**) and carbon dioxide; this process is termed **fermentation**.

The exchange of oxygen and carbon dioxide between body tissues and the environment is called 'external respiration', or ventilation. In air-breathing

vertebrates the exchange takes place in the alveoli of the lungs, aided by the muscular movements of breathing. Respiration at the cellular level is termed internal respiration, and in all higher organisms occurs in the mitochondria of the cells.

retrovirus

Any of a family of **viruses** (Retroviridae) containing **RNA** as the genetic material rather than the more usual **DNA**. For the virus to express itself and multiply within an infected cell, its RNA must be converted to DNA. It does this by using a built-in **enzyme** known as reverse transcriptase (since the transfer of genetic information from DNA to RNA is known as transcription, and retroviruses do the reverse of this). Retroviruses include those causing AIDS and some forms of leukaemia.

reverberation

The multiple reflections, or **echoes**, of sounds inside a building that merge and persist a short time (up to a few seconds) before fading away. At each **reflection** some of the **sound** energy is absorbed, causing the amplitude of the sound **wave** and the intensity of the sound to reduce a little. Too much reverberation causes sounds to become confused and indistinct, and this is particularly noticeable in empty rooms and halls, and such buildings as churches and cathedrals where the hard, unfurnished surfaces do not absorb sound **energy** well. Where walls and surfaces absorb sound energy very efficiently, too little reverberation may cause a room or hall to sound dull or 'dead'.

RNA (ribonucleic acid)

The nucleic acid involved in the process of translating the genetic material **DNA** into **proteins**. It is usually single-stranded, unlike the double-stranded DNA, and consists of a large number of nucleotides strung together, each of which carries one of four bases (uracil, cytosine, adenine, or guanine). RNA is copied from DNA by the formation of base pairs, with uracil taking the place of thymine.

RNA occurs in three major forms, each with a different function in the synthesis of protein molecules:

- *Messenger RNA* (mRNA) acts as the template for protein synthesis. Each codon (a set of three bases) on the RNA molecule is matched up with the corresponding **amino acid**, in accordance with the **genetic code**. This process (translation) takes place in the ribosomes, which are made up of proteins and *ribosomal RNA* (rRNA).

- *Transfer RNA* (tRNA) is responsible for combining with specific amino acids, and then matching up a special 'anticodon' sequence of its own with a codon on the mRNA. This is how the genetic code is translated.

Röntgen, Wilhelm Konrad (1845–1923)

German physicist who discovered **X-rays** in 1895 while investigating the passage of **electricity** through **gases**. The **radiation** he discovered in the process passed through some substances opaque to light, and affected photographic plates. Developments from this discovery revolutionized medical diagnosis. Röntgen won the Nobel Prize for Physics in 1901. He refused to make any financial gain out of his findings, believing that the products of scientific research should be made freely available to all.

Rutherford, Ernest, 1st Baron Rutherford of Nelson (1871–1937)

New Zealand-born British physicist who was a pioneer of modern atomic science. His main research was in the field of **radioactivity**. He discovered alpha, beta, and gamma rays and was the first to recognize the nuclear nature of the **atom** in 1911. Rutherford produced the first artificial transformation of one **element** into another in 1919, bombarding nitrogen with alpha particles and getting hydrogen and **oxygen**. After further research he announced that the **nucleus** of any atom must be composed of hydrogen nuclei; at Rutherford's suggestion, the name **'proton'** was given to the hydrogen nucleus in 1920. He speculated that uncharged particles (**neutrons**) must also exist in the nucleus. He was awarded a Nobel prize in 1908.

S

salt

A salt may be produced by chemical **reaction** between an **acid** and a base, or by the displacement of hydrogen from an **acid** by a **metal**. As a **solid**, the **ions** normally adopt a regular arrangement to form **crystals**. Some salts only form stable crystals as hydrates (when combined with water). Most inorganic salts readily dissolve in water to give an electrolyte (a **solution** that conducts **electricity**). Common table salt is sodium chloride, formed from the reaction between sodium hydroxide and hydrochloric acid.

Schrödinger, Erwin (1887–1961)

Austrian physicist who produced in 1926 a solid mathematical explanation of the **quantum theory** and the structure of the **atom**. He was awarded the Nobel prize in 1933.

In 1924 French physicist Louis de Broglie, using ideas from Albert **Einstein's** special theory of **relativity**, showed that an **electron** or any other particle has a wave associated with it. In 1926 both Schrödinger and de Broglie published the same wave equation, which Schrödinger later formulated in terms of the energies of the **electron** and the field in which it was situated. He solved the equation for the hydrogen atom and found that it fitted with energy levels proposed by Danish physicist Niels **Bohr**. In the hydrogen atom, the wave function describes where we can expect to find the electron. The electron does not follow a circular orbit, rather its position is described by the more complicated notion of an orbital, a region in space where the electron can be found with varying degrees of probability.

Schrödinger's cat

Schrödinger once proposed a paradoxical thought experiment to highlight one of the stranger aspects of quantum theory. Put a cat in a box, he said, along with a vial of poison and a sample of some radioactive element. There is a also a device that will be triggered to break open the poison vial when an atom of the radioactive material decays and emits a particle. After a certain period of time, depending on the element used, there's a fifty-fifty chance that an atom will decay, releasing the poison, and killing the cat.

There is, of course, an equal chance that the atom will not decay, and the cat will be spared. According to quantum mechanics, the atom is simultaneously in the decayed and undecayed states. Not until someone makes a measurement of the atom is it forced into one mode or the other. According to Schrödinger one would have to 'express this situation by having the living and the dead cat mixed, or smeared out (pardon the expression) into equal parts,' living and dead.

science

Any systematic field of study or body of knowledge that aims, through experiment, observation, and deduction, to produce reliable explanations of phenomena, with reference to the material and physical world.

Science is divided into separate areas of study, such as **astronomy, biology, geology, chemistry, physics**, and mathematics, although more recently attempts have been made to combine traditionally separate disciplines under such headings as life sciences and earth sciences. These areas are usually jointly referred to as the natural sciences. The physical sciences comprise mathematics, physics, and chemistry. The application of science for practical purposes is called technology.

scientific law

Principles that are taken to be universally applicable. Laws form the basic theoretical structure of the physical sciences, so that the rejection of a law by the scientific community is an almost inconceivable event. On occasion a law may be modified, as was the case when **Einstein** showed that **Newton's laws of motion** do not apply to objects travelling at speeds close to that of light.

season

A period of the year with a characteristic climate. The change in seasons is mainly due to the change in attitude of the Earth's axis in relation to the Sun, and hence the position of the Sun in the sky at a particular place. In temperate latitudes four seasons are recognized: spring, summer, autumn (fall), and winter. The northern temperate latitudes have summer when the southern temperate latitudes have winter, and vice versa. Tropical regions have two seasons – the wet and the dry. Monsoon areas around the Indian Ocean have three seasons: the cold, the hot, and the rainy.

seed

The reproductive structure of higher plants (**angiosperms** and **gymnosperms**). It develops from a fertilized ovule and consists of an **embryo** and a **food** store, surrounded and protected by an outer seed coat, called the testa. The food store is contained either in a specialized nutritive tissue, the endosperm, or in the cotyledons (seed leaves) of the embryo itself. In angiosperms the seed is enclosed within a **fruit**, whereas in gymnosperms it is usually naked and unprotected, once shed from the female cone.

seed *The structure of seeds. The castor is a dicotyledon, a plant in which the developing plant has two leaves, developed from the cotyledon. In maize, a monocotyledon, there is a single leaf developed from the scutellum.*

semiconductor

A material with electrical conductivity intermediate between **metals** and insulators and used in a wide range of electronic devices. Certain crystalline materials, most notably silicon and germanium, have a small number of free **electrons** that have escaped from the **bonds** between the **atoms**. The atoms from which they have escaped possess vacancies, called holes, which are similarly able to move from atom to atom and can be regarded as positive charges. Current can be carried by both electrons (negative carriers) and holes (positive carriers).

The conductivity of a semiconductor can be enhanced by doping the material with small numbers of atoms that either release free electrons (making an n-type semiconductor with more electrons than holes) or capture them (a p-type semiconductor with more holes than electrons). When p-type and n-type materials are brought together to form a p–n junction, an electrical barrier is formed which conducts current more readily in one direction than the other. This is the basis of the semiconductor diode and numerous other devices including transistors, rectifiers, and integrated circuits (silicon chips).

shock wave

Shock waves are produced when an object moves through a fluid at a supersonic speed. A body moving with supersonic speed cannot send signals through the atmosphere ahead of it; the disturbances that it creates can move away only sideways or to the rear. The forward limit of disturbances, the shock wave, is a very precisely defined boundary. Upstream of a shock wave the airflow is always supersonic. As the air passes through the shock wave it experiences an essentially instantaneous rise in **pressure**, **density**, and **temperature**. **Energy** is transferred to the air via the wave, and this results in a considerable increase in the backwards drag force experienced as a solid body accelerates through the speed of sound. Passage of a shock wave is heard as a sharp crack (small amplitude, as from a whip), a sudden bang (from a pistol), or one or more heavy booms (large amplitude, from thunder or an aeroplane).

singularity

General **relativity** predicts that a singularity will exist at the centre of a **black hole**. This is the point where the **density** of the material inside the hole becomes infinite and all known laws of time and space break down. However, since the singularity is hidden insides the hole's **event horizon** it has no effect on the outside **universe**. It is also thought, according to the **Big Bang** model of the origin of the universe, that a singularity was the starting point from which the expansion of the universe began.

SI units (French Système International d'Unités)

A standard system of scientific units, originally proposed in 1960, that is used by scientists worldwide.

Basic SI units:
- the metre (m) for length
- kilogram (kg) for mass
- second (s) for time
- ampere (A) for electrical current
- kelvin (K) for temperature
- mole (mol) for amount of substance
- candela (cd) for luminosity.

skeleton

The rigid or semi-rigid framework that supports and gives form to an animal's body, protects its internal organs, and provides anchorage points for its muscles. A skeleton may be internal, forming an endoskeleton, as in vertebrates, or external, forming an exoskeleton, as in the shells of insects or crabs. The skeleton may be composed of bone and cartilage (**vertebrates**), chitin (**arthropods**), calcium carbonate (**molluscs** and other **invertebrates**), or silica (many **protists**).

> **BONES**
>
> The human skeleton is composed of 206 bones, with the vertebral column (spine) forming the central supporting structure.

skin

The protective covering of the body of a **vertebrate**. it helps to protect the body from infection and to prevent dehydration. In **mammals**, the outer layer (epidermis) is dead and its cells are constantly being rubbed away and replaced from below. The lower layer (dermis) contains blood vessels, nerves, hair roots, and sweat and sebaceous glands, and is supported by a network of fibrous and elastic cells.

solid

A **state of matter** that holds its own shape (as opposed to a **liquid,** which takes up the shape of its container, or a **gas**, which totally fills its container). According to **kinetic theory**, the **atoms** or **molecules** in a solid are not free to move but merely vibrate about fixed positions.

solstice

Either of the days on which the Sun is farthest north or south of the celestial equator (an imaginary line projected on to the sky from the Earth's equator) each year. The summer solstice, when the Sun is farthest north, occurs around 21 June; the winter solstice around 22 December.

solution

A homogeneous mixture of a **liquid** with a **gas** or **solid**. One of the substances (usually the liquid) is the solvent and the others (solutes) are said to be dissolved in it. The solvent is normally the substance that is present in greatest quantity; however, if one of the constituents is a liquid this is considered to be the solvent even if it is not the major substance.

sound

The physiological sensation received by the ear, originating in a **vibration** that communicates itself as a **pressure** variation in the air and travels in every direction, spreading out as an expanding sphere. Like other **waves** – light waves and water waves – sound can be reflected, diffracted, and refracted. **Reflection** of a sound wave is heard as an **echo**. **Diffraction** explains why sound can be heard round doorways.

> **SPEED OF SOUND**
>
> All sound waves in air travel with a speed dependent on the **temperature**; under ordinary conditions, this is about 330 m/ 1,070 ft per second.

The pitch of the sound depends on the number of vibrations imposed on the air per second (**frequency**), but the speed is unaffected. The loudness of a sound is dependent primarily on the amplitude of the vibration of the air.

> **LOWS AND HIGHS**
>
> The lowest note audible to a human being has a frequency of about 20 hertz (vibrations per second), and the highest one of about 20,000 Hz; the lower limit of this range varies little with the person's age, but the upper range falls steadily from adolescence onwards.

Decibel scale

Decibels	Typical sound	Decibels	Typical sound
0	threshold of hearing	65–90	train
10	rustle of leaves in gentle breeze	75–80	factory (light/medium work)
10	quiet whisper	90	heavy traffic
20	average whisper	90–100	thunder
20–50	quiet conversation	110–140	jet aircraft at take-off
40–45	theatre (between performances)	130	threshold of pain
50–65	loud conversation	140–190	space rocket at take-off
65–70	traffic on busy street		

space-time

When developing **relativity**, Albert **Einstein** showed that time was in many respects like an extra dimension (or direction) to space. Space and time can thus be considered as entwined into a single entity, rather than two separate things. Space-time is considered to have four dimensions: three of space and one of time. In relativity theory, events are described as occurring at points in space-time. The general theory of relativity describes how space-time is distorted by the presence of material bodies, an effect that we observe as **gravity**.

species

A distinguishable group of organisms that resemble each other or consist of a few distinctive types (as in polymorphism), and that can all interbreed to produce fertile offspring. Species are the lowest level in the system of biological **classification**. Related species are grouped together in a genus. Within a species there are usually two or more separate **populations**, which may in time become distinctive enough to be designated subspecies or varieties, and could eventually give rise to new species through speciation.

Around 1.4 million species have been identified so far

- 750,000 are **insects**
- 250,000 are **plants**
- 41,000 are **vertebrates**.

spectrum (plural spectra)

A distribution of wavelengths or other properties arranged in order of magnitude. Visible light is part of the **electromagnetic spectrum**, which ranges from long wavelength **radio** waves to very short wavelength **gamma** rays. White **light** can be separated into red, orange, yellow, green, blue, indigo, and violet. The visible spectrum was first studied by Isaac **Newton**, who showed in 1672 how white light could be broken up into different **colours**.

spectroscopy

The study of spectra associated with atoms or molecules in solid, liquid, or gaseous phase. Spectroscopy can be used to identify unknown compounds and is an invaluable tool in science, medicine, and industry (for example, in checking the purity of drugs).

SOME TYPES OF SPECTROSCOPY

- Emission spectroscopy is the study of the characteristic series of sharp lines in the spectrum produced when an element is heated. Thus an unknown mixture can be analysed for its component elements.
- Absorption spectroscopy makes use of the fact that atoms and molecules absorb energy in a characteristic way, e.g. absorption bands in the infrared region of the spectrum are measured by infrared spectroscopy.
- Nuclear magnetic resonance (NMR) spectroscopy is concerned with interactions between adjacent atomic nuclei.

speed of light

The speed at which light and other electromagnetic waves travel through empty space. Its value is 299,792,458 m/186,281 mi per second. The speed of light is the highest speed possible, according to the theory of **relativity**, and its value is independent of the motion of its source and of the observer. It is impossible to accelerate any material body to this speed because it would require an infinite amount of **energy**.

standard model

The modern theory of elementary particles and their interactions. According to the standard model, elementary particles are classified as leptons (light particles, such as **electrons**), hadrons (particles, such as **neutrons** and **protons**, that are formed from **quarks**), and gauge bosons. Leptons and hadrons interact by exchanging gauge bosons, each of which is responsible for a different **fundamental force**: photons mediate the electromagnetic force, which affects all charged particles; gluons mediate the strong nuclear force, which affects quarks; gravitons mediate the force of **gravity**; and the weakons (intermediate vector bosons) mediate the weak nuclear force.

standard temperature and pressure (STP)

A standard set of conditions for experimental measurements, to enable comparisons to be made between sets of results.

Standard temperature: 0°C/32°F (273 K)
Standard pressure: 1 atmosphere (101,325 Pa).

states of matter

The forms (**solid**, **liquid**, or **gas**) in which matter can exist. Whether a material is solid, liquid, or gaseous depends on its **temperature** and the **pressure** on it. The transition between states takes place at definite temperatures, called **melting point** and **boiling point**. **Kinetic theory** describes how the state of a material depends on the movement and arrangement of its **atoms** or **molecules**.

- *Gases*: the atoms or molecules of gases move randomly in otherwise empty space, filling any size or shape of container. Gases can be liquefied by cooling as this lowers the speed of the molecules and enables attractive forces between them to bind them together.

- *Liquids* form a level surface and assume the shape of the container; the atoms or molecules do not occupy fixed positions, nor do they have total freedom of movement.

- *Solids* hold their own shape as their atoms or molecules are not free to move about but merely vibrate about fixed positions, such as those in crystal lattices.

A hot ionized gas or **plasma** is often called the fourth state of matter, but liquid crystals, colloids, and glass also have a claim to this title.

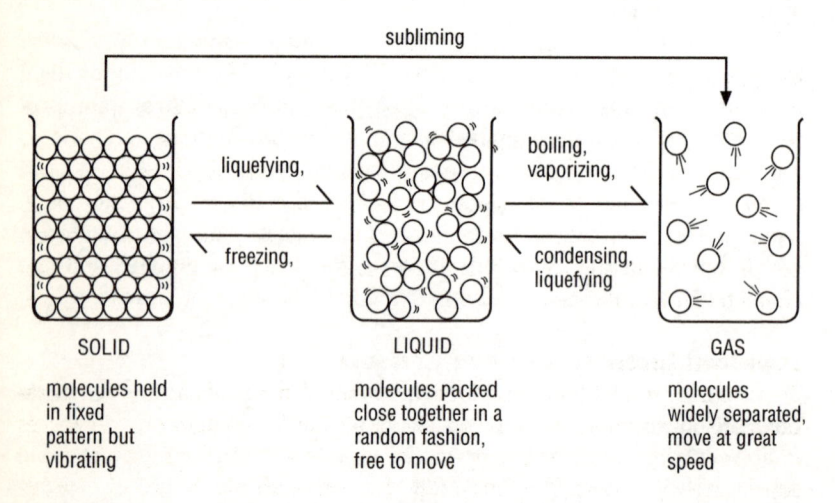

SOLID — molecules held in fixed pattern but vibrating

LIQUID — molecules packed close together in a random fashion, free to move

GAS — molecules widely separated, move at great speed

Transitions: liquefying, freezing (solid ↔ liquid); boiling, vaporizing, condensing, liquefying (liquid ↔ gas); subliming (solid → gas).

states of matter *The state (solid, liquid, or gas) of any substance is not fixed but varies with changes in temperature and pressure.*

static electricity
A stationary **electric charge**, usually acquired by a body by means of electrostatic induction or **friction**. Rubbing different materials can produce static electricity, as seen in the sparks produced on combing one's hair or removing a nylon shirt. In some processes static electricity is useful, as in paint spraying where the parts to be sprayed are charged with electricity of opposite polarity to that on the paint droplets.

stress and strain
Measures of the deforming **force** applied to a body (stress) and of the resulting change in its shape (strain). For a perfectly elastic material, stress is proportional to strain.

subatomic particle
A particle that is smaller than an **atom**. Such particles may be indivisible elementary particles, such as the **electron** and **quark**, or they may be composites, such as the **proton, neutron**, and alpha particle.

See also: *particle physics.*

substrate
A compound or mixture of compounds acted on by an enzyme. The term also refers to a substance such as agar that provides the nutrients for the metabolism of micro-organisms.

succession
A series of changes that occur in the structure and composition of the vegetation in a given area from the time it is first colonized by **plants** (primary succession), or after it has been disturbed by fire, flood, or clearing (secondary succession). If allowed to proceed undisturbed, succession leads naturally to a stable climax community (for example, oak and hickory forest or savannah grassland) that is determined by the climate and soil characteristics of the area.

superconductivity
The increase in electrical conductivity in some materials at low **temperatures**. The resistance of some **metals** and metallic **compounds** decreases uniformly with decreasing temperature until at a critical temperature (the superconducting point), within a few degrees of **absolute zero**, the **resistance** suddenly falls to zero. The phenomenon was discovered by Dutch scientist Heike Kamerlingh Onnes (1853–1926) in 1911.

superstring theory

A mathematical theory developed in the 1980s to explain the properties of elementary particles and the forces between them (in particular, **gravity** and the nuclear forces) in a way that combines **relativity** and **quantum theory**. In string theory, the fundamental objects in the universe are not pointlike particles but extremely small stringlike objects. These objects exist in a universe of ten dimensions, although, for reasons not yet understood, only three space dimensions and one dimension of time are discernible. There are many unresolved difficulties with superstring theory, but some physicists think it may be the ultimate 'theory of everything' that explains all aspects of the **universe** within one framework.

supersymmetry

A theory that relates the two classes of elementary particle, the fermions and the bosons (see **standard theory**). According to supersymmetry, each fermion particle has a boson partner particle, and vice versa. It has not been possible to marry up all the known fermions with the known bosons, and so the theory postulates the existence of other, as yet undiscovered fermions, such as the photinos (partners of the photons), gluinos (partners of the gluons), and gravitinos (partners of the gravitons). Using these ideas, it has become possible to develop a theory of gravity called supergravity that extends Einstein's work and considers the gravitational, nuclear, and electromagnetic forces to be manifestations of an underlying superforce. Supersymmetry has been incorporated into the **superstring** theory, and appears to be a crucial ingredient in the 'theory of everything' sought by scientists.

surface tension

The property that causes the surface of a **liquid** to behave as if it were covered with a weak elastic skin; this is why a needle can float on water. It is caused by the exposed surface's tendency to contract to the smallest possible area because of cohesive forces between **molecules** at the surface. Allied phenomena include the formation of droplets, the concave profile of a meniscus, and the capillary action by which water soaks into a sponge.

suspension

A mixture consisting of small solid particles dispersed in a **liquid** or **gas**, which will settle on standing. An example is milk of magnesia, which is a suspension of magnesium hydroxide in water.

symbiosis
Any close relationship between two organisms of different species where both partners benefit from the association. A well-known example is the **pollination** relationship between **insects** and **flowers**, where the insects feed on nectar and carry pollen from one flower to another. This is sometimes known as mutualism.

T

temperature
The degree or intensity of **heat** of an object and the condition that determines whether it will transfer heat to another object or receive heat from it, according to the laws of **thermodynamics**. The temperature of an object is a measure of the average kinetic energy possessed by the atoms or molecules of which it is composed. The SI unit of temperature is the kelvin (symbol K) used with the Kelvin scale (zero Kelvin = **absolute zero**).

terminal velocity
The maximum **velocity** that can be reached by a given object moving through a fluid (gas or liquid) under the action of an applied **force**. As the speed of the object increases so does the total magnitude of the forces resisting its motion. Terminal velocity is reached when the resistive forces exactly balance the applied force that has caused the object to accelerate; because there is now no resultant force, there can be no further acceleration. For example, an object falling through air will reach a terminal velocity and cease to accelerate under the influence of gravity when the air resistance equals the object's weight.

territory
A fixed area that an **animal** or group of animals will defend against other members of the same species. Animals may hold territories for many different reasons; for example, to provide a constant food supply, to monopolize potential mates, or to ensure access to refuges or nest sites. The size of a territory depends in part on its function and on the size of the animal: some nesting and mating territories may be only a few square metres, whereas feeding territories may be as large as hundreds of square kilometres.

thermodynamics
The branch of **physics** dealing with the transformation of heat into and from other forms of energy. It is the basis of the study of the efficient working of engines, such as the steam and internal combustion engines.

The three laws of thermodynamics

1 energy can be neither created nor destroyed
2 it is impossible for an unaided self-acting machine to convey heat from one body to another at a higher temperature
3 it is impossible by any procedure, no matter how idealized, to reduce any system to the **absolute zero** of temperature (0 K/-273°C/-459°F) in a finite number of operations.

Thomson, J(oseph) J(ohn) (1856–1940)
English physicist who is famous for discovering the electron and for his research into the conduction of electricity through gases, for which he was awarded the 1906 Nobel Prize for Physics. He also received several other honours, including a knighthood in 1908 and the Order of Merit in 1912.

time
Time is that dimension of **space-time** that allows us to distinguish two different events that occur in the same place. It is the interval between two such events that forms the basis of the measurement of time. Formerly this measurement was based on the Earth's rotation on its axis (the day) and its orbit around the Sun (a year). Scientifically time is now determined by **atomic clocks**.

toxin
Any poison produced by another living organism (usually a bacterium) that can damage another organism. In vertebrates, toxins are broken down by **enzyme** action, mainly in the liver.

trace element
A chemical element necessary in minute quantities for the health of a plant or animal. For example, magnesium, which occurs in chlorophyll, is essential to **photosynthesis**, and iodine is needed by the thyroid gland of mammals for making **hormones** that control growth and body chemistry.

transpiration
The loss of water from a plant by evaporation. Most water is lost from the leaves through

WATER LOSS

A single maize plant has been estimated to transpire 245 l/54 gal of water in one growing season.

pores known as stomata, whose primary function is to allow gas exchange between the plant's internal tissues and the atmosphere. Transpiration from the leaf surfaces causes a continuous upward flow of water from the roots known as the transpiration stream.

tropism (or tropic movement)

The directional growth of a plant, or part of a plant, in response to an external stimulus such as gravity or light. If the movement is directed towards the stimulus it is described as positive; if away from it, it is negative.

- *Geotropism*: response of plants to gravity, causing the root (positively geotropic) to grow downwards, and the stem (negatively geotropic) to grow upwards.
- *Phototropism*: response to light causing plants to grow towards light.
- *Hydrotropism*: plant response to water.
- *Chemotropism*: plant response to a chemical stimulus.
- *Thigmotropism,* or *haptotropism:* response to physical contact, as in the tendrils of climbing plants when they touch a support and then grow around it.

U

ultrasound

Pressure waves similar in nature to sound waves but occurring at **frequencies** above 20,000 Hz (cycles per second), the approximate upper limit of human hearing. Ultrasonics is concerned with the study and practical application of these phenomena. The earliest practical application was to detect submarines during World War I, but recently the field of ultrasonics has greatly expanded. Frequencies above 80,000 Hz have been used to produce echoes as a means of measuring the depth of the sea or to detect flaws in metal, and in medicine, high-frequency pressure waves are used to investigate various body organs. High-power ultrasound has been used with focusing arrangements to destroy deep-lying tissue in the body, and extremely high frequencies of 1,000 MHz (megahertz) or more are used in ultrasonic microscopes.

ultraviolet radiation

Electromagnetic radiation invisible to the human eye, of wavelengths from about 400 to 4 nm (where the **X-ray** range begins). Exposure to high levels of ultraviolet radiation causes sunburn and may trigger the growth of skin cancers. Most of the Sun's ultraviolet radiation is blocked by the **ozone** layer. Ultraviolet rays are strongly germicidal and may be produced artificially by mercury vapour and arc lamps for therapeutic use. The radiation may be detected with ordinary photographic plates or films. It can also be studied by its fluorescent effect on certain materials.

SEEING ULTRAVIOLET

Some animals can detect ultraviolet. It is sometimes known as bee purple from the ability of bees and other insects to detect ultraviolet patterns on flower petals. The desert iguana *Disposaurus dorsalis* uses it to locate the boundaries of its territory and to find food.

uncertainty principle
The principle that it is impossible to know with unlimited accuracy both the position and momentum of a particle. The principle arises because in order to locate a particle exactly, an observer must bounce light (in the form of a **photon**) off the particle, which must alter its position in an unpredictable way. It was established by German physicist Werner **Heisenberg**, and gave a theoretical limit to the precision with which a particle's momentum and position can be measured simultaneously: the more accurately one is determined, the more uncertainty there is in the other.

universal time (UT)
Another name for Greenwich Mean Time. It is based on the rotation of the Earth, which is not quite constant. Since 1972, UT has been replaced by coordinated universal time (UTC), which is based on uniform atomic time.

See also: *time.*

universe
All of space and everything it contains, the study of which is called **cosmology**. The universe is thought to be between 10 billion and 20 billion years old, and is mostly empty space, dotted with galaxies for as far as telescopes can see. The most distant detected galaxies and quasars lie 13 billion light years or more from Earth, and are moving farther apart as the universe expands. Several theories attempt to explain how the universe came into being and evolved; for example, the **Big Bang** theory of an expanding universe originating in a single event, and the contradictory steady-state theory.

V

vaccine
A weakened or killed form of a disease-causing bacteria or virus that is deliberately introduced into the body, usually either orally or by a hypodermic syringe, to induce the specific **antibody** reaction that produces **immunity** against the disease. In 1796, Edward **Jenner** was the first to inoculate a child successfully with cowpox virus to produce immunity to smallpox.

vacuum
A region of space in which there are very few atoms or molecules. In physics, any enclosure in which the gas pressure is considerably less than atmospheric pressure. A perfect vacuum, in which there would be no atoms and molecules at all, would be impossible to attain.

valency
The measure of an **element's** ability to combine with other elements, expressed as the number of atoms of hydrogen (or any other standard univalent element) capable of uniting with (or replacing) its atoms. The number of **electrons** in the outermost shell of the atom dictates the combining ability of an element.

Elements are described as uni-, di-, tri-, and tetravalent when they unite with one, two, three, and four univalent atoms respectively. Some elements have variable valency: for example, nitrogen and phosphorus have a valency of both three and five.

vapour
One of the three states of matter (see also **solid** and **liquid**). The molecules in a vapour move randomly and are far apart, the distance between them, and therefore the volume of the vapour, being limited only by the walls of any vessel in which they might be contained. A vapour differs from a gas only in that a vapour can be liquefied by increased pressure, whereas a gas cannot unless its temperature is lowered below its critical temperature; it then becomes a vapour and may be liquefied.

variation

A difference between individuals of the same species, found in any sexually reproducing population. Variations may be almost unnoticeable in some cases, obvious in others, and can concern many aspects of the organism. Typically, variations in size, behaviour, biochemistry, or colouring may be found. The cause of the variation is genetic (that is, inherited), environmental, or more usually a combination of the two. The origins of variation can be traced to the recombination of the genetic material during the formation of the gametes (sex cells), and, more rarely, to mutation.

velocity

The speed of an object in a given direction. Velocity is a vector quantity, since its direction is important as well as its magnitude (or speed). The velocity *v* of an object travelling in a fixed direction may be calculated by dividing the distance *s* it has travelled by the time *t* taken to do so, and may be expressed as:

$$v = \frac{s}{t}$$

vertebrate

Any animal with a backbone. The 41,000 species of vertebrates include mammals, birds, reptiles, amphibians, and fishes. In terms of the total number of species the vertebrates are only a tiny proportion of the world's animals.

FIRST VERTEBRATE

A giant fossil conodont (an eel-like organism) was discovered in South Africa in 1995, and is believed to be one of the first vertebrates. Conodonts evolved 520 million years ago, predating the earliest fish by about 50 million years.

---

virus

An infectious particle consisting of a core of nucleic acid (**DNA** or **RNA**) enclosed in a protein shell. Viruses are able to function and reproduce only if they can invade a living cell to use the cell's system to replicate themselves. In the process they may disrupt or alter the host cell's own DNA. Outside a living cell a virus is inert and to all intents and purposes lifeless.

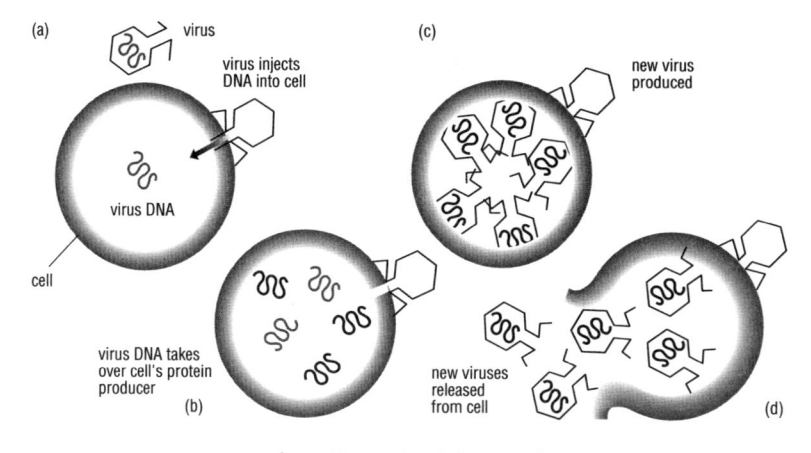

virus *How a virus infects a cell.*

Many viruses mutate continuously so that the host has little chance of developing permanent resistance; others transfer between species, with the new host similarly unable to develop resistance. The viruses that cause AIDS and Lassa fever are both thought to have 'jumped' to humans from other mammalian hosts. Among diseases caused by viruses are canine distemper, chickenpox, common cold, herpes, influenza, rabies, smallpox, yellow fever, AIDS, and many plant diseases. Recent evidence implicates viruses in the development of some forms of cancer (see **oncogenes**).

viscosity

The resistance of a fluid to flow, caused by its internal friction, which makes it resist flowing past a solid surface or other layers of the fluid. It applies to the motion of an object moving through a fluid as well as the motion of a fluid passing by an object. Fluids such as pitch, treacle, and heavy oils are highly viscous.

vitamin

Any of a number of organic compounds that are necessary in small quantities for the normal functioning of living organisms. Many act as coenzymes, small molecules that enable **enzymes** to function effectively. Vitamins must be supplied by the diet because the human body cannot make them. They are normally present in adequate amounts in a balanced diet. Deficiency of a vitamin may lead to a metabolic disorder ('deficiency disease'), which can be remedied by sufficient intake of the vitamin.

Volta, Alessandro Giuseppe Antonio Anastasio, Count (1745–1827)

Italian physicist who invented the first **electric cell** (the voltaic pile) in 1800. Volta made the first accurate estimate of the proportion of oxygen in the air by exploding air with hydrogen to remove the oxygen. Volta repeated and built on Italian physiologist Luigi **Galvani's** experiments with 'animal electricity' and in 1792, he concluded that the source of the electricity was the metals Galvani used and not, as Galvani thought, the animal tissue. Volta even succeeded in producing a list of metals in order of their electricity production based on the strength of the sensation they made on his tongue.

In 1800 Volta described two arrangements of conductors that produced an electric current. One was a pile of silver and zinc discs separated by cardboard moistened with brine, and the other a series of glasses of salty or alkaline water in which bimetallic curved electrodes were dipped. Volta's electric cell was a sensation, for it enabled high electric currents to be produced for the first time.

W

Wallace, Alfred Russel (1823–1913)
Welsh naturalist who collected animal and plant specimens in South America and South-East Asia, and independently arrived at a theory of evolution by natural selection similar to that proposed by Charles **Darwin**. In 1858, Wallace wrote an essay outlining his ideas on evolution and sent it to Darwin, who had not yet published his. Together they presented a paper to the Linnaean Society that year. Wallace's section, entitled 'On the Tendency of Varieties to Depart Indefinitely from the Original Type', described the survival of the fittest.

water
Water is the most common compound on Earth and vital to all living organisms. It covers 70% of the Earth's surface, and provides a habitat for large numbers of aquatic organisms. It is the largest constituent of all living organisms – the human body consists of about 65% water. It is chemical compound formed from the elements hydrogen and oxygen, H_2O. Pure water is a colourless, odourless, tasteless liquid which freezes at 0°C/32°F, and boils at 100°C/212°F.

WHERE THE WATER IS
Some 97% of the Earth's water is in the oceans; a further 2% is in the form of snow or ice, leaving only 1% available as fresh water for plants and animals.

- Water has a slight positive charge at one end and a slight negative charge at the other. The negatively charged oxygen atom attracts the positively charged hydrogen atoms of other water molecules, with the result that hydrogen bonds are formed between the water molecules holding them together. This makes water a very good solvent for other polar molecules and ionic substances.

- Water has a high heat capacity, which means that it requires large amounts of heat energy to produce small rises in temperature. Consequently, temperature changes in water are usually quite small, and

this is important in cells where metabolic reactions are controlled by enzymes
- A great deal of heat is required to change water from its liquid state to vapour and this is important in temperature control in mammals. When the body becomes overheated, the animal sweats; thus the heat used for the vaporization of water in sweat is lost from the body, thereby cooling it.
- When frozen, water expands by $\frac{1}{11}$ of its volume. It also become less dense so that ice floats on the surface.

Watson, James Dewey (1928–)

US biologist whose research on the molecular structure of **DNA** and the genetic code, in collaboration with Francis **Crick**, earned him a shared Nobel prize in 1962. Based on earlier works by scientists such as Rosalind **Franklin**, they were able to show that DNA formed a double helix of two spiral strands held together by base pairs.

WATSON'S LIFE	
1928	Watson is born on 6 April in Chicago, USA.
1943	At the age of just 15, he enters the University of Chicago to study zoology.
1947	He graduates at the age of 19.
1951	Watson goes to the Cavendish Laboratory at Cambridge University
1953	He publishes his work on the structure of DNA with fellow scientist Francis Crick.
1961	He is appointed Professor of Biology at Harvard University.
1962	He shares the Nobel Prize for Physiology or Medicine with Crick, for their work on DNA.
1968–1993	Watson is Director of the Cold Spring Harbor Laboratory of Quantitative Biology. He becomes President in 1994.
1989–1992	He is head of the US government's Human Genome Project.

wave

A regular series of disturbances travelling through space or a medium from a source. Waves carry energy but they do not transfer matter. There are two types of wave: in a longitudinal wave, such as a sound wave, the disturbance is parallel to the wave's direction of travel; in a transverse wave, such as an electromagnetic wave, it is perpendicular. The medium through which a wave travels (for example the Earth, for seismic waves) is not permanently displaced by the passage of a wave.

A longitudinal wave consists of a series of compressions and rarefactions (states of maximum and minimum density and pressure, respectively). Such waves are always mechanical in nature and thus require a medium through which to travel. Sound waves are an example of longitudinal waves.

- *Wavelength* is measured as the distance between successive crests (or successive troughs) of the wave.
- The *frequency* of a wave is the number of vibrations per second.
- The *speed* of the wave is measured by multiplying wave frequency by the wavelength.

  See also: *reflection, refraction.*

weight

The force exerted on an object by **gravity**. The weight of an object depends on its mass – the amount of material in it – and the strength of the gravitational pull acting on it. An object weighs less at the top of a mountain than at sea level because it is further from the centre of the Earth. On the surface of the Moon, an object has only one-sixth of its weight on Earth, because the Moon's surface gravity is one-sixth that of the Earth.

worm

Any one of a variety of elongated limbless invertebrates including the flatworms, such as flukes and tapeworms; the roundworms or nematodes, such as the eelworm and the hookworm; the marine ribbon worms or nemerteans; and the segmented worms or annelids.

ICE WORM

A new species of polychaete worm was discovered in 1997 on the floor of the Gulf of Mexico. It is a pink centipedelike worm that is about 5 cm/2 in long and lives in highly populated colonies in methane ice.

X-ray

The band of electromagnetic radiation in the wavelength range 10^{-11} to 10^{-9} m (between gamma rays and ultraviolet radiation; see **electromagnetic spectrum**). Applications of X-rays make use of their short wavelength (as in X-ray diffraction) or their penetrating power (as in medical X-rays of internal body tissues). X-rays are dangerous and can cause cancer.

X-rays with short wavelengths pass through most body tissues, although dense areas such as bone prevent their passage, showing up as white areas on X-ray photographs. The X-rays used in radiotherapy have very short wavelengths that penetrate tissues deeply and destroy them.

X-rays were discovered by German experimental physicist Wilhelm **Röntgen** in 1895.

Y

Young, Thomas (1773–1829)

English physicist, physician, and Egyptologist who revived the wave theory of light and identified the phenomenon of **interference** in 1801. He also established many important concepts in mechanics. In 1793, Young recognized that focusing of the eye (accommodation) is achieved by a change of shape in the lens of the eye, the lens being composed of muscle fibres. He also showed that astigmatism is due to irregular curvature of the cornea. In 1801, he became the first to recognize that colour sensation is due to the presence in the retina of structures that respond to the three colours red, green, and violet.

YOUNG'S LIFE

1773 Young is born on 13 June in Milverton, Somerset, UK.

1800 He opens a medical practice in London.

1801 He is appointed Professor of Natural Philosophy at the Royal Institute, but resigns from this post two years later. He becomes the first to identify the structures in the eye that give colour sensation and announces his major discovery of the principle of interference in light waves.

1807 Young establishes many important concepts in mechanics through a published series of lectures.

1815 He starts to publish a series of papers on Egyptology and is one of the first to interpret the writings on the Rosetta Stone.

1817 He suggests that light waves may contain a transverse component, this is proved in 1821.

1829 Young dies in London on 10 May.

zoology
The branch of biology concerned with the study of animals. It includes any aspect of the study of animal form and function, the study of evolution of animal forms, anatomy, physiology, embryology, behaviour, and geographical distribution.

Appendix

Amino Acids

Name	Formula
glycine	$CH_2(NH_2).COOH$
alanine	$CH_3CH.(NH_2).COOH$
phenylalanine	$C_6H_5CH_2CH.(NH_2).COOH$
tyrosine	$C_6H_4OH.CH_2CH.(NH_2).COOH$
valine	$(CH_3)_2CH.CH.(NH_2).COOH$
leucine	$(CH_3)_2CH.CH_2CH.(NH_2).COOH$
iso-leucine	$(CH_3).CH_2CH(CH_3)CH.(NH_2).COOH$
serine	$CH_2OH.CH.(NH_2).COOH$
threonine	$CH_3CHOH.CH.(NH_2).COOH$
cysteine	$SH.CH_2CH.(NH_2).COOH$
methionine	$CH_3.S.(CH_2)_2CH.(NH_2).COOH$
asparagine	$NH_2CO.CH_2CH.(NH_2).COOH$
glutamine	$NH_2CH.(CH_2)_2(CO.NH_2).COOH$
lysine	$NH_2CH_3CH.(NH_2)$
arginine	$NH_2C(NH).NH(CH_2)_3CH.(NH_2).COOH$
aspartic acid	$COOH.CH_2CH.(NH_2).COOH$
glutamic acid	$COOH.(CH_2)_2CH.(NH_2).COOH$
histidine	$C_3H_3N_2.CH_2CH.(NH_2).COOH$
trytophan	$C_4.NH.CH_2CH_2CH.(NH_2).COOH$
proline	$NH.(CH_2)_3CH.COOH$

Biochemistry: Chronology

c. 1830 Johannes Müller discovers proteins.

1833 Anselme Payen and J F Persoz first isolate an enzyme.

1862 Haemoglobin is first crystallized.

1869 The genetic material DNA (deoxyribonucleic acid) is discovered by Friedrich Mieschler

1899 Emil Fischer postulates the "lock-and-key" hypothesis to explain the specificity of enzyme action.

1913 Leonor Michaelis and M L Menten develops a mathematical equation describing the rate of enzyme-catalysed reactions.

1915 The hormone thyroxine is first isolated from thyroid gland tissue.

1920 The chromosome theory of heredity is postulated by Thomas H Morgan; growth hormone is discovered by Herbert McLean Evans and J A Long.

1921 Insulin is first isolated from the pancreas by Frederick Banting and Charles Best.

1926	Insulin is obtained in pure crystalline form.
1927	Thyroxine is first synthesized.
1928	Alexander Fleming discovers penicillin.
1931	Paul Karrer deduces the structure of retinol (vitamin A); vitamin D compounds are obtained in crystalline form by Adolf Windaus and Askew, independently of each other.
1932	Charles Glen King isolates ascorbic acid (vitamin C).
1933	Tadeus Reichstein synthesizes ascorbic acid.
1935	Richard Kuhn and Karrer establish the structure of riboflavin (vitamin B2).
1936	Robert Williams establishes the structure of thiamine (vitamin B1); biotin is isolated by Kogl and Tonnis.
1937	Niacin is isolated and identified by Conrad Arnold Elvehjem.
1938	Pyridoxine (vitamin B6) is isolated in pure crystalline form.
1939	The structure of pyridoxine is determined by Kuhn.
1940	Hans Krebs proposes the Krebs (citric acid) cycle; Hickman isolates retinol in pure crystalline form; Williams establishes the structure of pantothenic acid; biotin is identified by Albert Szent-Györgyi, Vincent Du Vigneaud, and co-workers.
1941	Penicillin is isolated and characterized by Howard Florey and Ernst Chain.
1943	The role of DNA in genetic inheritance is first demonstrated by Oswald Avery, Colin MacLeod, and Maclyn McCarty.
1950	The basic components of DNA are established by Erwin Chargaff; the alpha-helical structure of proteins is established by Linus Pauling and R B Corey.
1953	James Watson and Francis Crick determine the molecular structure of DNA.
1956	Mahlon Hoagland and Paul Zamecnick discover transfer RNA (ribonucleic acid); mechanisms for the biosynthesis of RNA and DNA are discovered by Arthur Kornberg and Severo Ochoa.
1957	Interferon is discovered by Alick Isaacs and Jean Lindemann.
1958	The structure of RNA is determined.
1960	Messenger RNA is discovered by Sidney Brenner and François Jacob.
1961	Marshall Nirenberg and Ochoa determines the chemical nature of the genetic code.
1965	Insulin is first synthesized.
1966	The immobilization of enzymes is achieved by Chibata.
1968	Brain hormones are discovered by Roger Guillemin and Andrew Schally.
1975	J Hughes and Hans Kosterlitz discover encephalins.
1976	Guillemin discovers endorphins.
1977	J Baxter determines the genetic code for human growth hormone.
1978	Human insulin is first produced by genetic engineering.
1979	The biosynthetic production of human growth hormone is announced by Howard Goodman and J Baxter of the University of California, and by D V Goeddel and Seeburg of Genentech.

1982 Louis Chedid and Michael Sela develop the first synthesized vaccine.
1983 The first commercially available product of genetic engineering (Humulin) is launched.
1985 Alec Jeffreys devises genetic fingerprinting.
1993 UK researchers introduce a healthy version of the gene for cystic fibrosis into the lungs of mice with induced cystic fibrosis, restoring normal function.
1996 Japanese chemists successfully synthesize cellulose.
1997 US geneticists construct the first artificial human chromosome.

Biology: Chronology

c. 500 BC First studies of the structure and behaviour of animals, by the Greek Alcmaeon of Croton.
c. 450 Hippocrates of Kos undertakes the first detailed studies of human anatomy.
c. 350 Aristotle lays down the basic philosophy of the biological sciences and outlines a theory of evolution.
c. 300 Theophrastus carries out the first detailed studies of plants.
c. AD 175 Galen establishes the basic principles of anatomy and physiology.
c. 1500 Leonardo da Vinci studies human anatomy to improve his drawing ability and produces detailed anatomical drawings.
1628 William Harvey describes the circulation of the blood and the function of the heart as a pump.
1665 Robert Hooke uses a microscope to describe the cellular structure of plants.
1672 Marcelle Malphigi undertakes the first studies in embryology by describing the development of a chicken egg.
1677 Anton van Leeuwenhoek greatly improves the microscope and uses it to describe spermatozoa as well as many micro-organisms.
1736 Carolus (Carl) Linnaeus publishes his systematic classification of plants, so establishing taxonomy.
1768–79 James Cook's voyages of discovery in the Pacific reveal a great diversity of living species, prompting the development of theories to explain their origin.
1796 Edward Jenner establishes the practice of vaccination against smallpox, laying the foundations for theories of antibodies and immune reactions.
1809 Jean-Baptiste Lamarck advocates a theory of evolution through inheritance of acquired characteristics.
1839 Theodor Schwann proposes that all living matter is made up of cells.
1857 Louis Pasteur establishes that micro-organisms are responsible for fermentation, creating the discipline of microbiology.
1859 Charles Darwin publishes *On the Origin of Species*, expounding his theory of the evolution of species by natural selection.

1865	Gregor Mendel pioneers the study of inheritance with his experiments on peas, but achieves little recognition.
1883	August Weismann proposes his theory of the continuity of the germ plasm.
1900	Mendel's work is rediscovered and the science of genetics founded.
1935	Konrad Lorenz publishes the first of many major studies of animal behaviour, which founds the discipline of ethology.
1953	James Watson and Francis Crick describe the molecular structure of the genetic material DNA.
1964	William Hamilton recognizes the importance of inclusive fitness, so paving the way for the development of sociobiology.
1975	Discovery of endogenous opiates (the brain's own painkillers) opens up a new phase in the study of brain chemistry.
1976	Har Gobind Khorana and his colleagues construct the first artificial gene to function naturally when inserted into a bacterial cell, a major step in genetic engineering.
1982	Gene databases are established at Heidelberg, Germany, for the European Molecular Biology Laboratory, and at Los Alamos, USA, for the US National Laboratories.
1985	The first human cancer gene, retinoblastoma, is isolated by researchers at the Massachusetts Eye and Ear Infirmary and the Whitehead Institute, Massachusetts.
1988	The Human Genome Organization (HUGO) is established in Washington, DC, with the aim of mapping the complete sequence of DNA.
1991	Biosphere 2, an experiment that attempts to reproduce the world's biosphere in miniature within a sealed glass dome, is launched in Arizona, USA.
1992	Researchers at the University of California, USA, stimulate the multiplication of isolated brain cells of mice, overturning the axiom that mammalian brains cannot produce replacement cells once birth has taken place. The world's largest organism, a honey fungus with underground hyphae (filaments) spreading across 600 hectares/1,480 acres, is discovered in Washington State, USA.
1994	Scientists from Pakistan and the USA unearth a 50-million-year-old fossil whale with hind legs that would have enabled it to walk on land.
1995	New phylum identified and named Cycliophora. It contains a single known species, *Symbion pandora*, a parasite of the lobster.
1996	The sequencing of the genome of brewer's yeast *Saccharomyces cerevisiae* is completed, the first time this has been achieved for an organism more complex than a bacterium. The 12 million base pairs took 300 scientists six years to map. A new muscle is discovered by two US dentists. It is 3 cm/1 in long, and runs from the jaw to behind the eye socket.
1997	The first mammal to be cloned from a nonreproductive cell is born. The lamb (named Dolly) has been cloned from an udder cell from a six-year-old ewe.
1999	Canadian researchers engineer an artificial chromosome that can be inserted into mammal cells and then transmitted from one generation to the next. The development has important implications towards germ-line therapy where a defect is corrected in the gametes and the change passed on to future generations.

Chemistry: Chronology

c. 3000 BC	Egyptians are producing bronze – an alloy of copper and tin.
c. 450 BC	Greek philosopher Empedocles proposes that all substances are made up of a combination of four elements – earth, air, fire, and water – an idea that is developed by Plato and Aristotle and persists for over 2,000 years.
c. 400 BC	Greek philosopher Democritus theorizes that matter consists ultimately of tiny, indivisible particles, atomos.
AD 1	Gold, silver, copper, lead, iron, tin, and mercury are known.
200	The techniques of solution, filtration, and distillation are known.
7th–17th centuries	Chemistry is dominated by alchemy, the attempt to transform nonprecious metals such as lead and copper into gold. Though misguided, it leads to the discovery of many new chemicals and techniques, such as sublimation and distillation.
12th century	Alcohol is first distilled in Europe.
1242	Gunpowder is introduced to Europe from the Far East.
1620	The scientific method of reasoning is expounded by Francis Bacon in his *Novum Organum*.
1650	Leyden University in the Netherlands sets up the first chemistry laboratory.
1661	Robert Boyle defines an element as any substance that cannot be broken down into still simpler substances and asserts that matter is composed of 'corpuscles' (atoms) of various sorts and sizes, capable of arranging themselves into groups, each of which constitutes a chemical substance.
1662	Boyle describes the inverse relationship between the volume and pressure of a fixed mass of gas (Boyle's law).
1697	Georg Stahl proposes the erroneous theory that substances burn because they are rich in a substance called phlogiston.
1755	Joseph Black discovers carbon dioxide.
1774	Joseph Priestley discovers oxygen, which he calls 'dephlogisticated air'. Antoine Lavoisier demonstrates his law of conservation of mass.
1777	Lavoisier shows air to be made up of a mixture of gases, and shows that one of these – oxygen – is the substance necessary for combustion (burning) and rusting to take place.
1781	Henry Cavendish shows water to be a compound.
1792	Alessandra Volta demonstrates the electrochemical series.
1807	Humphry Davy passes an electric current through molten compounds (the process of electrolysis) in order to isolate elements, such as potassium, that have never been separated by chemical means. Jöns Berzelius proposes that chemicals produced by living creatures should be termed 'organic'.
1808	John Dalton publishes his atomic theory, which states that every element consists of similar indivisible particles – called atoms – which differ from the atoms of other

elements in their mass; he also draws up a list of relative atomic masses. Joseph Gay-Lussac announces that the volumes of gases that combine chemically with one another are in simple ratios.

1811 Amedeo Avogadro's hypothesis on the relation between the volume and number of molecules of a gas, and its temperature and pressure, is published.

1813–14 Berzelius devises the chemical symbols and formulae still used to represent elements and compounds.

1828 Franz Wöhler converts ammonium cyanate into urea – the first synthesis of an organic compound from an inorganic substance.

1832–33 Michael Faraday expounds the laws of electrolysis, and adopts the term 'ion' for the particles believed to be responsible for carrying current.

1846 Thomas Graham expounds his law of diffusion.

1853 Robert Bunsen invents the Bunsen burner.

1858 Stanislao Cannizzaro differentiates between atomic and molecular weights (masses).

1861 Organic chemistry is defined by German chemist Friedrich Kekulé as the chemistry of carbon compounds.

1864 John Newlands devises the first periodic table of the elements.

1869 Dmitri Mendeleyev expounds his periodic table of the elements (based on atomic mass), leaving gaps for elements that have not yet been discovered.

1874 Jacobus van't Hoff suggests that the four bonds of carbon are arranged tetrahedrally, and that carbon compounds can therefore be three-dimensional and asymmetric.

1884 Swedish chemist Svante Arrhenius suggests that electrolytes (solutions or molten compounds that conduct electricity) dissociate into ions, atoms or groups of atoms that carry a positive or negative charge.

1894 William Ramsey and Lord Rayleigh discover the first inert gas, argon.

1897 The electron is discovered by J J Thomson.

1901 Mikhail Tsvet invents paper chromatography as a means of separating pigments.

1909 Sören Sörensen devises the pH scale of acidity.

1912 Max von Laue shows crystals to be composed of regular, repeating arrays of atoms by studying the patterns in which they diffract X-rays.

1913–14 Henry Moseley equates the atomic number of an element with the positive charge on its nuclei, and draws up the periodic table, based on atomic number, that is used today.

1916 Gilbert Newton Lewis explains covalent bonding between atoms as a sharing of electrons.

1927 Nevil Sidgwick publishes his theory of valency, based on the numbers of electrons in the outer shells of the reacting atoms.

1930 Electrophoresis, which separates particles in suspension in an electric field, is invented by Arne Tiselius.

1932 Deuterium (heavy hydrogen), an isotope of hydrogen, is discovered by Harold Urey.

1940 Edwin McMillan and Philip Abelson show that new elements with a higher atomic number than uranium can be formed by bombarding uranium with neutrons, and synthesize the first transuranic element, neptunium.

1942 Plutonium is first synthesized by Glenn T Seaborg and Edwin McMillan.

1950 Derek Barton deduces that some properties of organic compounds are affected by the orientation of their functional groups (the study of which becomes known as conformational analysis).

1954 Einsteinium and fermium are synthesized.

1955 Ilya Prigogine describes the thermodynamics of irreversible processes (the transformations of energy that take place in, for example, many reactions within living cells).

1962 Neil Bartlett prepares the first compound of an inert gas, xenon hexafluoroplatinate; it was previously believed that inert gases could not take part in a chemical reaction.

1965 Robert B Woodward synthesizes complex organic compounds.

1981 Quantum mechanics is applied to predict the course of chemical reactions by US chemist Roald Hoffmann and Kenichi Fukui of Japan.

1982 Element 109, unnilennium, is synthesized.

1985 Fullerenes, a new class of carbon solids made up of closed cages of carbon atoms, are discovered by Harold Kroto and David Walton at the University of Sussex, England.

1987 US chemists Donald Cram and Charles Pederson, and Jean-Marie Lehn of France create artificial molecules that mimic the vital chemical reactions of life processes.

1990 Jean-Marie Lehn, Ulrich Koert, and Margaret Harding report the synthesis of a new class of compounds, called nucleohelicates, that mimic the double helical structure of DNA, turned inside out.

1993 US chemists at the University of California and the Scripps Institute synthesize rapamycin, one of a group of complex, naturally occurring antibiotics and immunosuppressants that are being tested as anticancer agents.

1994 Elements 110 (ununnilium) and 111 (unununium) are discovered at the GSI heavy-ion cyclotron, Darmstadt, Germany.

1995 German chemists build the largest ever wheel molecule, made up of 154 molybdenum atoms surrounded by oxygen atoms. It has a relative molecular mass of 24,000 and is soluble in water.

1996 Element 112 is discovered at the GSI heavy-ion cyclotron, Darmstadt, Germany.

1997 The International Union of Pure and Applied Chemistry (IUPAC) states that elements 104–109 should be named rutherfordium (104), dubnium (105), seaborgium (106), bohrium (107), hassium (108), and meitnerium (109).

1999 Russian scientists at the Institute of Nuclear Research at Dubna create element 114 by colliding isotopes calcium 48 and plutonium 44. Shortly afterwards, US physicists create element 118, which decays into another new element, 116, by bombarding lead with krypton.

Cloning: Chronology

1975 British scientist Derek Brownhall produces the first clone of a rabbit, in Oxford, UK.

1981 Chinese scientists make the first clone of a fish (a golden carp).

1984 Allan Wilson and Russell Higuchi of the University of California, Berkeley, USA, clone genes from an extinct animal, the quagga. Sheep are successfully cloned.

1988 The first dairy cattle are produced by cloning embryos.

1996 US geneticists clone two rhesus monkeys from embryo cells.

1997 British geneticists clone an adult sheep. A cell is taken from the udder of the mother sheep and its DNA (deoxyribonucleic acid) is combined with an unfertilized egg that has had its DNA removed. The fused cells are grown in the laboratory and then implanted into the uterus of a surrogate mother sheep. The resulting lamb, Dolly, comes from an animal that is six years old. This is the first time cloning has been achieved using cells other than reproductive cells. The news is met with international calls to prevent the cloning of humans.

US president Bill Clinton announces a ban on using federal funds to support human cloning research, and calls for a moratorium on this type of scientific research. He also asks the National Bioethics Advisory Commission to review and issue a report on the ramifications that cloning will have on humans.

US genetic scientist Don Wolf announces the production of monkeys cloned from embryos. It is a step closer to cloning humans and raises acute philosophical issues.

1998 Doctors meeting at the World Medical Association's conference in Hamburg, Germany, call for a worldwide ban on human cloning. US president Clinton calls for legislation banning cloning the following day.

Dolly, the sheep who was cloned in 1997, gives birth to a female lamb at the Roslin Institute in Edinburgh, Scotland.

1999 Dolly is discovered to have mitochondrion from the egg cell and so it is not an exact clone.

Genetics: Chronology

1856 Austrian monk and botanist Gregor Mendel begins experiments breeding peas that will lead him to the laws of heredity.

1865 Gregor Mendel publishes a paper in the *Proceedings of the Natural Science Society of Brünn* that outlines the fundamental laws of heredity.

1869 Swiss biochemist Johann Miescher discovers a nitrogen and phosphorous material in cell nuclei that he calls nuclein but which is now known as the genetic material DNA.

1888 Dutch geneticist Hugo Marie de Vries uses the term 'mutation' to describe varieties that arise spontaneously in cultivated primroses.

1902	US geneticist Walter Sutton and German zoologist Theodor Boveri find the chromosomal theory of inheritance when they show that cell division is connected with heredity.
1906	English biologist William Bateson introduces the term 'genetics'.
1910	US geneticist Thomas Hunt Morgan discovers that certain inherited characteristics of the fruit fly *Drosophila melanogaster* are sex linked. He later argues that because all sex-related characteristics are inherited together they are linearly arranged on the X chromosome.
1934	Norwegian biochemist Asbjrn Fölling discovers the genetic metabolic defect phenylketonuria, which can cause retardation; his discovery stimulates research in biochemical genetics and the development of screening tests for carriers of deleterious genes.
1944	The role of deoxyribonucleic acid (DNA) in genetic inheritance is first demonstrated by US bacteriologist Oswald Avery, US biologist Colin MacLeod, and US biologist Maclyn McCarthy; it opens the door to the elucidation of the genetic code.
1945	Working in Japan, US geneticist Samuel G Salmon discovers *Norin 10*, a semidwarf wheat variety which grows quickly, responds well to fertilizer, does not fall over from the weight of the grains, and, when crossed with disease-resistant strains in the USA, results in a wheat strain that increases wheat harvests by more than 60% in India and Pakistan.
25 April 1953	English molecular biologist Francis Crick and US biologist James Watson announce the discovery of the double helix structure of DNA, the basic material of heredity. They also theorize that if the strands are separated then each can form the template for the synthesis of an identical DNA molecule. It is perhaps the most important discovery in biology.
1954	Russian-born US cosmologist George Gamow suggests that the genetic code consists of the order of nucleotide triplets in the DNA molecule.
1958	US geneticists George Beadle, Edward Tatum, and Joshua Lederberg share the Nobel Prize for Physiology or Medicine: Beadle and Tatum for their discovery that genes act by regulating definite chemical events; and Lederberg for his discoveries concerning genetic recombination.
1961	French biochemists François Jacob and Jacques Monod discover messenger ribonucleic acid (mRNA), which transfers genetic information to the ribosomes, where proteins are synthesized.
1967	US scientist Charles Caskey and associates demonstrate that identical forms of messenger RNA produce the same amino acids in a variety of living beings, showing that the genetic code is common to all life forms.
1967	US biochemist Marshall Nirenberg establishes that mammals, amphibians, and bacteria all share a common genetic code.
1968	US geneticists Mark Ptashne and Walter Gilbert separately identify the first repressor genes.

1969 US geneticist Jonathan Beckwith and associates at Harvard Medical School isolate a single gene for the first time.

The Nobel Prize for Physiology or Medicine is awarded jointly to US physiologists Max Delbrück, Alfred Hershey, and Salvador Luria for their discoveries concerning the replication mechanism and genetic structure of viruses.

1970 US geneticist Hamilton Smith discovers type II restriction enzyme that breaks the DNA strand at predictable places, making it an invaluable tool in recombinant DNA technology.

US biochemists Howard Temin and David Baltimore separately discover the enzyme reverse transcriptase, which allows some cancer viruses to transfer their RNA to the DNA of their hosts turning them cancerous – a reversal of the common pattern in which genetic information always passes from DNA to RNA.

1972 US microbiologist Daniel Nathans uses a restriction enzyme that splits DNA molecules to produce a genetic map of the monkey virus (SV40), the simplest virus known to produce cancer; it is the first application of these enzymes to an understanding of the molecular basis of cancer.

Venezuelan-born US immunologist Baruj Benacerraf and Hugh O'Neill McDevitt show immune response to be genetically determined.

1973 US biochemists Stanley Cohen and Herbert Boyer develop the technique of recombinant DNA. Strands of DNA are cut by restriction enzymes from one species and then inserted into the DNA of another; this marks the beginning of genetic engineering.

1975 The gel-transfer hybridization technique for the detection of specific DNA sequences is developed; it is a key development in genetic engineering.

1976 US biochemist Herbert Boyer and venture capitalist Robert Swanson found Genentech in San Francisco, California, the world's first genetic engineering company.

28 August: Indian-born US biochemist Har Gobind Khorana and his colleagues announce the construction of the first artificial gene to function naturally when inserted into a bacterial cell. This is a major breakthrough in genetic engineering.

1977 US scientist Herbert Boyer, of the firm Genentech, fuses a segment of human DNA into the bacterium *Escherichia coli* which begins to produce the human protein somatostatin; this is the first commercially produced genetically engineered product.

1980 A new vaccine for the prevention of hepatitis B is tested in the USA. It is the first genetically engineered vaccine and has a success rate of 92%. It wins Federal Drug Administration approval in 1986.

16 June: The US Supreme Court rules that a microbe created by genetic engineering can be patented.

1981 The US Food and Drug Administration grants permission to Eli Lilley and Co to market insulin produced by bacteria, the first genetically engineered product to go on sale.

The genetic code for the hepatitis B surface antigen is discovered, creating the possibility of a bioengineered vaccine.

1981 US geneticists Robert Weinberg, Geoffrey Cooper, and Michael Wigler discover that oncogenes (genes that cause cancer) are integrated into the genome of normal cells.

1982 Using genetically engineered bacteria, the Swedish firm Kabivitrum manufactures human growth hormone.

1983 Geneticist James Gusella identifies the gene for Huntington's disease.

1984 British geneticist Alec Jeffreys discovers that a core sequence of DNA is almost unique to each person; this examination of DNA, known as 'genetic fingerprinting', can be used in criminal investigations and to establish family relationships.

1986 The US Department of Agriculture permits the Biological Corporation of Omaha to market a virus produced by genetic engineering; it is the first living genetically altered organism to be sold. The virus is used against a form of swine herpes.

The US Department of Agriculture permits the first outdoor test of genetically altered high-yield plants (tobacco plants).

1987 German-born British geneticist Walter Bodmer and associates announce the discovery of a marker for a gene that causes cancer of the colon.

The first genetically altered bacteria are released into the environment in the USA; they protect crops against frost.

Foxes in Belgium are immunized against rabies by using bait containing a genetically engineered vaccine, dropped from helicopters. The success of the experiment leads to a large-scale vaccination programme.

April: The US Patent and Trademark Office announces its intention to allow the patenting of animals produced by genetic engineering.

10 October: The *New York Times* announces Dr Helen Donis-Keller's mapping of all 23 pairs of human chromosomes, allowing the location of specific genes for the prevention and treatment of genetic disorders.

1988 The US Patent and Trademark Office grants Harvard University a patent for a mouse developed by genetic engineering.

1989 Scientists in Britain introduce genetically engineered white blood cells into cancer patients, to attack tumours.

1991 British geneticists Peter Goodfellow and Robin Lovell-Badge discover the gene on the Y chromosome that determines sex.

1992 The US biotechnology company Agracetus patents transgenic cotton, which has had a foreign gene added to it by genetic engineering.

US biologist Philip Leder receives a patent for the first genetically engineered animal, the oncomouse, which is sensitive to carcinogens.

1993 US geneticist Dean Hammer and colleagues at the US National Cancer Institute publish the approximate location of a gene that could predispose human males to homosexuality.

1994 Trials using transfusions of artificial blood begin in the USA. The blood contains genetically engineered haemoglobin.

February: The US Food and Drug Administration approves the use of genetically engineered bovine somatotropin (BST), which increases a cow's milk yield by 10–40%. It is banned in Europe.

May: The first genetically engineered food goes on sale in California and Chicago, Illinois. The 'Flavr Savr' tomato is produced by the US biotechnology company Calgene.

1995 A genetically engineered potato is developed that contains the gene for Bt toxin, a natural pesticide produced by a soil bacterium. The potato plant produces Bt within its leaves.

US embryologists Edward Lewis and Eric Wieschaus and German embryologist Christiane Nüsslein-olhard are jointly awarded the Nobel Prize for Physiology or Medicine for their discoveries concerning the genetic control of early embryonic development.

Australian geneticists produce a genetically engineered variety of cotton that contains a gene from a soil bacteria that kills the cotton bollworm and native budworm.

Trials begin in the USA to treat breast cancer by gene therapy. The women are injected with a virus genetically engineered to destroy their tumours.

April: US surgeons report the successful transplant of genetically altered hearts of pigs into baboons, a notable advance in trans-species operations.

July: The US government approves experimentation of genetically altered animal organs in humans.

August: The US Environmental Protection Agency approves the sale of genetically modified maize, which contains a gene from a soil bacterium that produces a toxin fatal to the European corn borer, a pest that causes approximately $1 billion damages annually.

1996 The first genetically engineered salmon are hatched, at Loch Fyne in Scotland. The salmon contain genes from the ocean pout as well as a salmon growth hormone gene that causes them to grow five times as fast as other salmon.

9 May: Scientists at the National Institute of Allergy and Infectious Disease discover a protein, fusin, which allows the HIV virus to fuse with a human immune system cell's outer membrane and inject genetic material. Its presence is necessary for the AIDS virus to enter the cell.

August: US geneticists clone two rhesus monkeys from embryo cells.

1997 *27 February:* Scottish researcher Ian Wilmut of the Roslin Institute in Edinburgh, Scotland, announces that British geneticists have cloned an adult sheep. A cell was taken from the udder of the mother sheep and its DNA combined with an unfertilized egg that had had its DNA removed. The fused cells were grown in the laboratory and then implanted into the uterus of a surrogate mother sheep. The resulting lamb, Dolly, came from an animal that was six years old.

This is the first time cloning has been achieved using cells other than reproductive cells. The news is met with international calls to prevent the cloning of humans.

February: US genetic scientist Don Wolf announces the production of monkeys cloned from embryos.

It is a step closer to cloning humans and raises acute philosophical issues.

16 May: US geneticists identify a gene 'clock' in chromosome 5 in mice that regulates the circadian rhythm.

3 June: US geneticist Huntington F Wilard constructs the first artificial human chromosome. He inserts telomeres (which consist of DNA and protein on the tips of chromosomes) and centromeres (specialized regions of DNA within a chromosome) removed from white blood cells into human cancer cells which are then assembled into chromosomes which are about one-tenth the size of normal chromosomes. The artificial chromosome is successfully passed on to all daughter cells.

11 June: English behavioural scientist David Skuse claims that boys and girls differ genetically in the way they acquire social skills. Girls acquire social skills intuitively and are 'pre-programmed', while boys have to be taught. This has important implications for education.

August: US geneticist Craig Venter and colleagues publish the genome of the bacterium *Helicobacter pylori*, a bacterium that infects half the world's population and which is the leading cause of stomach ulcers. It is the sixth bacterium to have its genome published, but is the most clinically important. It has 1,603 putative genes, encoded in a single circular chromosome that is 1,667,867 nucleotide base-pairs of DNA long. Complete genomes are increasingly being published as gene-sequencing techniques improve.

18 September: US geneticist Bert Vogelstein and colleagues demonstrate that the p53 gene, which is activated by the presence of carcinogens, induces cells to commit suicide by stimulating them to produce large quantities of poisonous chemicals, called "reactive oxygen species" (ROS).

The cells literally poison themselves. It is perhaps the human body's most effective way of combating cancer. Many cancers consist of cells with a malfunctioning p53 gene.

November: The US Food and Drug Administration (FDA) approves Rituxan, the first anticancer monoclonal antibody made from genetically engineered mouse antibodies. The antibody binds itself to non-Hodgkin's lymphoma (a cancer of the lymph system) cancer cells and triggers the immune system to kill the cells.

1998 *October:* US scientist French Anderson announces a technique that could cure inherited diseases by inserting a healthy gene to replace a damaged one.

He calls for a full debate on the issue of gene therapy, which brings with it the dilemma of whether it is ethical to enable the choice of physical attributes such as eye colour and height.

8 December: The Human Fertilization and Embryology Authority and the Human Genetics Advisory Commission publish a joint report in the UK on cloning.

While they oppose cloning for reproductive purposes, they leave the door open for cloning for curing intractable diseases.

10 December: In a joint effort by scientists around the world, the first genetic blueprint for a whole multicellular animal is completed. The 97 million-letter code, which is published on the Internet, is for a tiny worm called *C elegans*. The study began 15 years ago and cost £30 million.

1999 *24 January:* US scientist Craig Venter of the Institute for Genomic Research in Maryland announces the possibility of creating a living, replicating organism from an artificial set of genes, at a meeting of the American Association for the Advancement of Science, in Anaheim, California. The experiment is put on hold until the moral question is discussed by religious leaders and ethicists at the University of Pennsylvania.

April: European Union legislation is implemented in the UK, requiring that some foods containing GM protein or DNA be labelled in restaurants and food shops.

18 May: A group of scientists at a specially convened Royal Society meeting finds that the experiments of Hungarian-born doctor Arpad Pusztai had on genetically modified foods were 'fundamentally flawed'. (In August 1999 Pusztai claimed that his experiments demonstrated that genetically modified potatoes stunted the growth of laboratory rats, strengthening public opinion in the UK against genetically modified foods.)

2000 First working draft of the Human Genome Project, which maps the composition and sequence of all human genes, is due to be released.

Nuclear Energy: Chronology

1896 French physicist Henri Becquerel discovers radioactivity.

1905 In Switzerland, Albert Einstein shows that mass can be converted into energy.

1911 New Zealand physicist Ernest Rutherford proposes the nuclear model of the atom.

1919 Rutherford splits the atom, by bombarding a nitrogen nucleus with alpha particles.

1939 Otto Hahn, Fritz Strassman, and Lise Meitner announce the discovery of nuclear fission.

1942 Enrico Fermi builds the first nuclear reactor, in a squash court at the University of Chicago, USA.

1946 The first fast reactor, called Clementine, is built at Los Alamos, New Mexico.

1951 The Experimental Breeder Reactor, Idaho, USA, produces the first electricity to be generated by nuclear energy.

1954 The first reactor for generating electricity is built in the USSR, at Obninsk.

1956 The world's first commercial nuclear power station, Calder Hall, comes into operation in the UK.

1957 The release of radiation from Windscale (now Sellafield) nuclear power station, Cumbria, UK, causes 39 deaths to 1977. In Kyshym, USSR, the venting of plutonium waste causes high but undisclosed casualties (30 small communities are deleted from maps produced in 1958).

1959	Experimental fast reactor built in Dounreay, northern Scotland.
1979	Nuclear-reactor accident at Three Mile Island, Pennsylvania, USA.
1986	An explosive leak from a reactor at Chernobyl, the Ukraine, results in clouds of radioactive material spreading as far as Sweden; 31 people are killed and thousands of square kilometres are contaminated.
1991	The first controlled and substantial production of nuclear-fusion energy (a two-second pulse of 1.7 megawatt) is achieved at JET, the Joint European Torus, at Culham, Oxfordshire, UK.
1994	The world record for producing nuclear fusion energy is broken at the Tokamak Fusion Test Reactor at Princeton University, New Jersey. The Princeton tokamak produces 9 megawatts of power.
1995	Sizewell B, the UK's first pressurized-water nuclear reactor and the most advanced nuclear power station in the world, begins operating in Suffolk, UK.
1996	British Nuclear Fuels is fined £25,000 after safety failures that result in a worker becoming contaminated with radioactivity.
1997	English physicists at JET (Joint European Torus) fuse isotopes of deuterium and tritium to produce a record 12 megawatts of nuclear-fusion power.
1999	A nuclear accident in Tokaimura, Japan, results in 320,000 people told to remain at home with all doors and windows closed.

Physics: Chronology

***c.* 400 BC**	The first 'atomic' theory is put forward by Democritus.
***c.* 250**	Archimedes' principle of buoyancy is established.
AD 1600	Magnetism is described by William Gilbert.
1608	Hans Lippershey invents the refracting telescope.
***c.* 1610**	The principle of falling bodies descending to earth at the same speed is established by Galileo.
1642	The principles of hydraulics are put forward by Blaise Pascal.
1643	The mercury barometer is invented by Evangelista Torricelli.
1656	The pendulum clock is invented by Christiaan Huygens.
1662	Boyle's law concerning the behaviour of gases is established by Robert Boyle.
***c.* 1665**	Isaac Newton puts forward the law of gravity, stating that the Earth exerts a constant force on falling bodies.
1690	The wave theory of light is propounded by Christiaan Huygens.
1704	The corpuscular theory of light is put forward by Isaac Newton.
1714	The mercury thermometer is invented by Daniel Fahrenheit.

1764	Specific and latent heats are described by Joseph Black.
1771	The link between nerve action and electricity is discovered by Luigi Galvani.
c. 1787	Charles's law relating the pressure, volume, and temperature of a gas is established by Jacques Charles.
1795	The metric system is adopted in France.
1798	The link between heat and friction is discovered by Benjamin Rumford.
1800	Alessandro Volta invents the Voltaic cell.
1801	Interference of light is discovered by Thomas Young.
1808	The 'modern' atomic theory is propounded by John Dalton.
1811	Avogadro's hypothesis relating volumes and numbers of molecules of gases is proposed by Amedeo Avogadro.
1814	Fraunhofer lines in the solar spectrum are mapped by Joseph von Fraunhofer.
1815	Refraction of light is explained by Augustin Fresnel.
1820	The discovery of electromagnetism is made by Hans Oersted.
1821	The dynamo principle is described by Michael Faraday; the thermocouple is invented by Thomas Seebeck.
1822	The laws of electrodynamics are established by André Ampère.
1824	Thermodynamics as a branch of physics is proposed by Sadi Carnot.
1827	Ohm's law of electrical resistance is established by Georg Ohm; Brownian movement resulting from molecular vibrations is observed by Robert Brown.
1829	The law of gaseous diffusion is established by Thomas Graham.
1831	Electromagnetic induction is discovered by Faraday.
1834	Faraday discovers self-induction.
1842	The principle of conservation of energy is observed by Julius von Mayer.
c. 1847	The mechanical equivalent of heat is described by James Joule.
1849	A measurement of speed of light is put forward by French physicist Armand Fizeau (1819–1896).
1851	The rotation of the Earth is demonstrated by Jean Foucault.
1858	The mirror galvanometer, an instrument for measuring small electric currents, is invented by William Thomson (Lord Kelvin).
1859	Spectrographic analysis is made by Robert Bunsen and Gustav Kirchhoff.
1861	Osmosis is discovered.
1873	Light is conceived as electromagnetic radiation by James Maxwell.
1877	A theory of sound as vibrations in an elastic medium is propounded by John Rayleigh.
1880	Piezoelectricity is discovered by Pierre Curie.
1887	The existence of radio waves is predicted by Heinrich Hertz.
1895	X-rays are discovered by Wilhelm Röntgen.
1896	The discovery of radioactivity is made by Antoine Becquerel.
1897	Joseph Thomson discovers the electron.

1899	Ernest Rutherford discovers alpha and beta rays.
1900	Quantum theory is propounded by Max Planck; the discovery of gamma rays is made by French physicist Paul-Ulrich Villard (1860–1934).
1902	Oliver Heaviside discovers the ionosphere.
1904	The theory of radioactivity is put forward by Rutherford and Frederick Soddy.
1905	Albert Einstein propounds his special theory of relativity.
1908	The Geiger counter is invented by Hans Geiger and Rutherford.
1911	The discovery of the atomic nucleus is made by Rutherford.
1913	The orbiting electron atomic theory is propounded by Danish physicist Niels Bohr.
1915	X-ray crystallography is discovered by William and Lawrence Bragg.
1916	Einstein puts forward his general theory of relativity; mass spectrography is discovered by William Aston.
1924	Edward Appleton makes his study of the Heaviside layer.
1926	Wave mechanics is introduced by Erwin Schrödinger.
1927	The uncertainty principle of quantum physics is established by Werner Heisenberg.
1931	The cyclotron is developed by Ernest Lawrence.
1932	The discovery of the neutron is made by James Chadwick; the electron microscope is developed by Vladimir Zworykin.
1933	The positron, the antiparticle of the electron, is discovered by Carl Anderson.
1934	Artificial radioactivity is developed by Frédéric and Irène Joliot-Curie.
1939	The discovery of nuclear fission is made by Otto Hahn and Fritz Strassmann.
1942	The first controlled nuclear chain reaction is achieved by Enrico Fermi.
1956	The neutrino, an elementary particle, is discovered by Clyde Cowan and Fred Reines.
1960	The Mössbauer effect of atom emissions is discovered by Rudolf Mössbauer; the first laser and the first maser are developed by US physicist Theodore Maiman (1927–).
1964	Murray Gell-Mann and George Zweig discover the quark.
1967	Jocelyn Bell (now Bell Burnell) and Antony Hewish discover pulsars (rapidly rotating neutron stars that emit pulses of energy).
1971	The theory of superconductivity is announced, where electrical resistance in some metals vanishes above absolute zero.
1979	The discovery of the asymmetry of elementary particles is made by US physicists James W Cronin and Val L Fitch.
1982	The discovery of processes involved in the evolution of stars is made by Subrahmanyan Chandrasekhar and William Fowler.
1983	Evidence of the existence of weakons (W and Z particles) is confirmed at CERN, validating the link between the weak nuclear force and the electromagnetic force.
1986	The first high-temperature superconductor is discovered, able to conduct electricity without resistance at a temperature of −238°C/−396°F.
1989	CERN's Large Electron Positron Collider (LEP), a particle accelerator with a circumference of 27 km/16.8 mi, comes into operation.

1991 LEP experiments demonstrate the existence of three generations of elementary particles, each with two quarks and two leptons.

1995 Top quark is discovered at Fermilab, the US particle-physics laboratory, near Chicago. US researchers announce the discovery of a material which is superconducting at the temperature of liquid nitrogen – a much higher temperature than previously achieved.

1996 CERN physicists create the first atoms of antimatter (nine atoms of antihydrogen). The Lawrence Livermore National Laboratory, California, USA, produces a laser of 1.3 petawatts (130 trillion watts).

1997 A new subatomic particle, an exotic meson, is possibly discovered at Brookhaven National Laboratory, Upton, New York, USA. The exotic meson is made up of either a quark, an antiquark and a gluon, or two quarks and two antiquarks. US physicists display the first atomic laser. It emits atoms that act like lightwaves.

1999 Scientists succeed in slowing down the speed of light from its normal speed of 299,792 km/186,282 mi per second to 61 km/38 mi per hour, opening up potential for the development of high-precision computer and telecommunications technologies, as well as for the advanced study of quantum mechanics.

Physical Constants

Physical constants, or fundamental constants, are standardized values whose parameters do not change.

Constant	Symbol	Value in SI units
acceleration of free fall	g	9.80665 m s^{-2}
Avogadro's constant	N_A	6.0221367×10^{23} mol^{-1}
Boltzmann's constant	k	1.380658×10^{-23} J K^{-1}
elementary charge	e	$1.60217733 \times 10^{-19}$ C
electronic rest mass	m_e	$9.1093897 \times 10^{-31}$ kg
Faraday's constant	F	9.6485309×10^{4} C mol^{-1}
gas constant	R	8.314510 J K^{-1} mol^{-1}
gravitational constant	G	6.672×10^{-11} N m^2 kg^{-2}
Loschmidt's number	N_L	2.686763×10^{25} m^{-3}
neutron rest mass	m_n	$1.6749286 \times 10^{-27}$ kg
Planck's constant	h	$6.6260755 \times 10^{-34}$ J s
proton rest mass	m_p	$1.6726231 \times 10^{-27}$ kg
speed of light in a vacuum	c	2.99792458×10^{8} m s^{-1}
standard atmosphere	atm	1.01325×10^{5} Pa
Stefan–Boltzmann constant	σ	5.67051×10^{-8} W m^{-2} K^{-4}1